Principles of
Organic Farming

Principles of Organic Farming

(For Degree students of Agriculture Universities)

L. P. Awasthi
Formerly, Dean, School of Agriculture
RNB Global University, Bikaner, Rajasthan, India

Siddhartha Das
The Institute of Plant Sciences
Dept. of Ornamental Plants and Agricultural Biotechnology
ARO, the Volcani Center, POB 15159
68 Hamakabim Road, 7528809, Rishon Lezion, Israel

Sudeepta Pattanayak
Division of Plant Pathology
ICAR-Indian Agricultural Research
Institute (IARI), New Delhi, India

TECHSAR™

NEW DELHI

Published and Printed by
Techsar Pvt. Ltd.
4435-36/7, Ansari Road, Daryaganj
New Delhi-110 002 (India)
E-mail: tech_info@techsarworld.com
Website: www.techsarworld.com

ISBN 978-93-90620-10-4

Reprint 2025

Foreword

Prof. A. N. Mukhopadhyay
Ph.D., D.Sc. FNAAS, HFIPS, HFMSI
Former Vice Chancellor AAU and
Director General Tea Res Association

Principles of Organic Farming (For Degree students of agriculture universities) describes key aspects of organic agriculture and provides a unique and science-based resource for researchers, teachers, extension workers, students, primary producers and others around the world.

Organic farming is an agricultural practice aimed at attaining a balanced ecological production management system that encourages and improves soil biological activity and biodiversity. It is a method of crop and livestock production that involves much more than choosing not to use pesticides, fertilizers, genetically modified organisms, antibiotics and growth hormones. It may provide attentive care that promotes the health and meets the behavioural needs of livestock. Organic farming system in India is not new and is being followed from ancient time. This farming system uses organic wastes in combination with other biological materials such as beneficial microbes to keep the soil alive and healthy. This noble approach can increase the agricultural production in a sustainable ecofriendly way. FAO suggested that "Organic agriculture is an unique production management system which promotes and enhances agro-ecosystem health, including biodiversity, biological cycles and soil biological activity, and this is accomplished by using on-farm agronomic, biological and mechanical methods in exclusion of all synthetic off-farm inputs". With the increase in population our compulsion would be not only to stabilize agricultural production but to increase it further in sustainable manner. The scientists have realized that the 'Green Revolution' with high input use has reached a plateau and is now sustained with diminishing return of falling dividends. Thus, a natural balance needs to be maintained at all cost for existence of life and property. The obvious choice for that would be more relevant in the present era, when these agrochemicals which are produced from fossil fuel and are not renewable and are diminishing in availability. It may also cost heavily on our foreign exchange in future.

The book consists of five theoretical units and one practical unit. Units one to five describe fundamentals of organic farming procedure, production requirement, government initiative, marketing strategy, role and operational structure of NPOP, certification procedure. Unit six or practical part consists of organic farm management, preparation of enriched compost, vermicomposting, biofertilizers/ bio-inoculants, ITK based nutrient, insect, pest disease and weed management strategies. All the contents exactly fulfil the basic of fundamental course curriculum of organic farming in agriculture.

The book gathers together a range of specialists with direct experience of organic farming over many years. This book is intended to be a unique and indispensable resource that offers a diverse range of valuable information, data and perspectives on organic agriculture at a time when the world community is increasingly aware of the problems of our current agricultural practices and the importance of creating sustainable agricultural and systems for the long-term health of humankind and the biosphere as a whole.

This book is a milestone in the new awareness and significance of organic farming. Prof. L.P. Awasthi, Formerly, Dean School of Agriculture at R.N.B. Global University, Bikaner, Dr. Siddhartha Das, The Institute of Plant Sciences, ARO, the Volcani Center, Israel and Dr. Sudeepta Pattanayak, Division of Plant Pathology, IARI, New Delhi, India, the authors of this book, are well-known Agriculture Scientists/Plant Pathologists and have done tremendous research work on Organic Agriculture. It covers every aspect of the subject in a readable style and in sufficient detail to enable students, researchers and professors to acquire and disseminate nature and role of organic farming.

I am sure this book will be very useful for students, researchers, professors and also farmers/entrepreneurs want to grow organic food.

(Prof. A.N. Mukhopadhyay)

Preface

Organic farming appealed to me because it involved searching for and discovering nature's pathways, as opposed to the formulaic approach chemical farming. The appeal of organic farming is boundless; this mountain has no top, this river has no end.

– *Eliot Coleman*

The present book is an attempt to provide graduate and postgraduate students with a technical tool, by which they can assist and establish themselves as future progressive agriculturists. Sustainable agriculture is implemented only when converting our chemical based agricultural system into organic based farming system. The book is framed based on organic farming courses of different agricultural universities and other universities of India. Discussion with the students, teachers and respective academic personnel has brought out the fact that a handy good quality, information enriched organic farming book is highly demanded. Therefore, we endeavoured to compile comprehensive information as well as a perfect syllabus topic oriented output with this book.

This is a compilation work and embodies a fairly comprehensive and crisp treatment of the fundamental facts and aspects of organic farming to prepare graduate and postgraduate students of Agriculture stream for vocational courses to achieve success in degree courses as well as different competitive examinations.

The book deals with the landmarks in the field of sustainable agriculture and organic farming, brief introduction, strategies of promoting organic farming, vermicomposting, green manuring, soil improvement and amendments, biofertilizers, weed management, bio-pesticides, certification process, etc. The book contains total five units which largely cover approximately sixteen chapters with addition of six practical based chapters. This book is strictly prepared as par ICAR-UGC graduate and postgraduate syllabus. At the end of each chapter, there is an important question series designed to provoke individual thought and critical examination of their memory to increase the conceptual and analytical view of a particular chapter. After completion of Unit I-V, a detailed practical session discusses by following of ICAR syllabus, which helps the budding progressive agriculturists to understand the subject and strengthen their bases.

We gratefully wish the students and various readers who enjoy working in the field of organic farming will find the format of book, level and quantity of information contained in the book to be appropriate. The wide spectrum of information in various chapters with the addition of the terms related to chapter and concept statements is presented in a concise manner. It is expected that the readers will find the book and conducive to learning and preparing themselves for degree and competitive exams with confidence.

We would like to thank our family and fellow authors who have helped and encourage us to complete this book. This book is dedicated to all organic farming warriors who directly or indirectly fight against chemical revolution in agriculture and promote organic farming for sustainable agriculture.

Prof. L.P. Awasthi
Dr. Siddhartha Das
Ms. Sudeepta Pattanayak

Contents

UNIT 3:
BIO-INTENSIVE NUTRIENT MANAGEMENT AND BIOFERTILIZERS

UNIT 5: QUALITY MAINTENANCE, CERTIFICATION AND MARKETING

Unit I

Concept of Organic Farming

1 CHAPTER
Concept and Relevance of Organic Farming

1.1 INTRODUCTION

Excessive use of agro-chemicals and their hazardous nature generates the need of alternative eco-friendly farming approaches in terms of organic management. Organic farming practices are traditionally come from our forefathers. Sustainable agriculture is important to reach the core of sustainable development. Organic farming is one of the crucial approaches which meets the objective of sustainable agriculture. Various techniques are applied in organic farming like mulching, inter-cropping, mixed cropping, integration of new crops, etc. Organic farming is associated with various laws and certification programmes, which prohibit the enormous use of synthetic agro-chemicals. Adverse effect of modern agricultural practices are on, habitat, human health and environmental health. Hazardous effects on the environment are manifested through soil toxicity and erosion, water scarcity and salinity, soil contamination and genetic erosion. Organic farming is one of the best methods to neutralize negative effects of toxic hazardous chemicals.

1.2 CONCEPTS OF ORGANIC FARMING

Organic farming is a productive system which totally prohibits the use of synthetic agro-chemicals and gives us eco-friendly management strategy. Agricultural practices rely to the maximum extent to the crop residues, crop rotation, mixed cropping, animal manures, mulching, biofertilizers, etc.

1.3 KINDS OF ORGANIC FARMING

1.3.1 Natural Farming

In this technique natural ingredients or components are used to make consortia or colloidal paste as natural plant growth enhancer. Two different components – bijamruta – for seed treatment and jivamruta – for soil treatment and foliar spray are found to be most effective.

1.3.2 Vedic Farming

In this system use of chemical pesticides, fertilizers and chemical growth regulators are restricted. It is the most ancient traditional farming system, which prefers the use of natural ingredients. Vedic farming provides healthy soils, healthy food and environment, sustainable farming land.

1.3.3 Homa Farming

These types of eco-friendly farming practices are very unique. It involves burning of rice grains with clarified butter in a pyramidal structure with mantra chanting. Through this traditional rituals, end material fly ash, is used as a good quality plant growth promoter.

1.3.4 Biodynamic Farming

In this system farm is considered a living system which interacts with various components like soil, environment, etc. Under this principle compost is manufactured from dead material. In this technique chemical fertilizers are replaced by microbial (algae, fungi, bacteria, mycorhiza, actinomycetes) fertilizers. Biological pest management is implemented through predators, parasites, other plethora or natural pest enemies. Various kinds of composting, green manuring, trap crops are the major tool under this mode of farming system.

1.4 ORGANIC VERSUS INORGANIC FARMING

Organic and inorganic agriculture are on two different poles. There are so many positive and negative impacts as we compare the two. Organic is the decentralization concept whereas inorganic is the centralization concept. Organic farming is independent in nature and inorganic farming is dependent on nature. Organic farming supports community whereas inorganic farming maintains competition. Organic farming is ecofriendly, non-toxic non-hazardous in nature, whereas inorganic farming is completely opposite. Diverse organic products are available in the market whereas inorganic products are very specific.

1.5 PRINCIPLES OF ORGANIC FARMING

Organic farming is an important eco-friendly approach. It has several dimensions:

- By using natural products in organic production it organizes along with livestock and the management of other resources.
- Appropriate technologies are crucially important under the biological system.

- Soil fertility is maintained through organic farming using renewable resources.
- Diversification is applied for optimum reduction.
- Increases the nutritional value of the food.
- Applying and use decentralized structures for distributing, marketing and processing of products.
- Sustainable ecofriendly approach and relationship between manmade activities and soil or land.
- Balance between wildlife and habitats through its multidirectional approach.

1.6 GLOBAL STATUS OF ORGANIC FARMING

Organic agriculture has travelled a long way and cosmopolitically distributed among 172 countries. Organic farming is a global phenomenon. It has a huge demand in almost all countries. Every year on an average 30-40% domestic organic markets grow. Investment in the company or industrial sector lowers the organic food products demand as well as demoralizes organic farming. Thailand is an organically improved country. Local farmers have practised traditional farming for long. Such practices have been developed and enriched through farmer's knowledge of local agro-ecology and environmentally sustainable ways of farming. In 1980s, many farmers and non-governmental organizations (NGOs) constructed an alternative agricultural network (AAN) to foster sustainable agriculture. In 2002, Organic Agriculture Certification Thailand (ACT) was established. In 2009, 135.44 million US dollar revenue was generated through organic farming. Countries like Mongolia, Netherlands are also taking progressive role in this aspect. Land use policy states that in Oceania (12.1 million hectares), Latin America (8.1 million hectares) and Europe (8.2 million hectares) was used for organic production. About one-third of the world's organically managed agricultural land, i.e., approximately 12 million hectares, is located in developing countries. The countries with the largest area under organic management are Argentina, China and Brazil—about two-thirds of the agricultural land. The total organic production area is of 3.3 million hectares. India and China are the two largest organic producers in Asia. India's organic production is near about 2 million hectares.

1.7 ORGANIC FARMING IN INDIA

India ranks second in population in the world hence, agriculture and food security is the main concern. Agricultural production has increased in the last few decades. Bhattacharyya and Chakroborty,

(2005) described current status of organic farming of India. India holds a prestigious position in organic production in global perspective. In 2009, revenue generated through organic production was $69 million and further estimated 12% to $77 million. Organic products demand seems to rise again post-COVID 19. In 2018-19 organic farming covered is 1.08 per cent of total agricultural area land production reached 1.93 million hectares, which actually increased by 8.8%, as compared to 2017. West Bengal, Maharashtra, Madhya Pradesh, Rajasthan, Uttar Pradesh, and Sikkim actively participate in organic farming and certification. Approximately, 453,622 farmers adopted organic farming practices. Singh in 2004, point out the main key issues of Indian farming. In 2018-19 organic production in India was 2.6 million metric tonnes. Different crops are introduced under this farming system like oilseeds, pulses, sugar crops, millets, vegetables, etc. Organic product investment per crop is represented in Table 1.1.

Table 1.1 Different crops and their organic product investment

Crop category	Organic	In conversion	Total
Vegetables	7134	1	7135
Fruits	35813	2	35815
Spices and condiments	56208	45	56253
Pulses	71875	0	71875
Sugar crops	984730	6910	991640
Oil seeds	727148	8	727156
Fibre crops	312945	888	313833

1.8 SIGNIFICANCE OF ORGANIC FARMING

Organic farming is significant in terms of:

- maintaining high nutritional quality of food and sufficient production,
- balancing socio-ecological impact on organic production,
- maintaining soil fertility on a long-term basis,
- maintaining genetic diversity of the production system and its surroundings including natural wildlife biodiversity,
- promoting proper care of waterbodies and underground water,
- recycling of renewable resources under localized production system,
- providing all livestock conditions of life that help to moderate their behaviour,
- lowering all kinds of pollution and maintaining ecological balance and sustainable land for future farming, and
- improving the production, processing and distribution chain.

As per latest statistics near about 30-35% agricultural needs are fulfilled by organic farming. Role of hazardous, toxic, non-ecofriendly inorganic fertilizers needs to be revised. Organic farming policies also need to be revised.

1.9 CONCLUSION

Organic farming gives us more safe and nutritious food. Day by day increased food demand makes the organic food market high. Organic farming is ecofriendly. It maintains soil health and environmental integrity. Organic import-export and new markets are growing faster. According to Willer and Lernoud, 2019 India is the largest organic producer and in near future will compete with any developing country through nation building in terms of ecological, nutritional and economical aspects.

1.10 QUESTIONS

Long Questions

1. Briefly describe the current status of organic farming globally and in India.
2. Write an essay – "Role of organic farming on global perspective".
3. Write a detailed note on the principles of organic farming. Mention its significance.

Short Questions

1. Describe different kinds of traditional organic farming strategies.
2. Compare organic and inorganic farming.
3. Significance of organic farming.
4. Write a short note on Indian organic farming.

REFERENCES

Bhattacharyya, P. and Chakroborty, G. (2005). Current status of organic farming in India and other countries. *Indian Journal Fertilizers*, 1: 118.

Singh, S.K. (2004). Organic farming : A key issue for Indian Agriculture. *Indian Farmers Digest*. 37(2): 5.

Willer H., Lernoud, J. (2019). The world organic agriculture statistics and emerging trends. Research Institute of Organic Agriculture, FiBL, IFOAM Organic International, available at http://www.organic-world.net/yearbook/yearbook-2019.html

2 CHAPTER

Initiatives Taken by Government (Central/ State), NGOs and Other Organizations for Promotion of Organic Agriculture

2.1 INTRODUCTION

Organic farming comes from very old traditional farming system and was the focus at the time of green revolution. Properly maintained farm reduces the soil and water pollution, and conserves the biodiversity. India stands in a good position for its geographical position with cultivated land and diverse agro-cultivated conditions. Day by day organic agriculture is gaining attention in agri-farming system. Development in agri-farming system promotes rural income. Government takes various initiatives through different schemes, order and projects. In this chapter we focus on different state and central government initiatives in Indian agri-farming system.

2.2 CENTRAL/STATE GOVERNMENT INITIATIVE TO PROMOTE ORGANIC FARMING IN INDIA

In 2010, through a special task force, Ministry of Agriculture, Government of India gave special importance to organic farming. The aforesaid task force helps in promoting of organic farming and shares its benefits for public awareness. In 1984, the first conference was held at Wardha on organic farming in India. The central and state governments, different civil societies and non-governmental organizations and concerned individuals promote organic agriculture. Indian Government took various initiatives in like the establishment of National Institute of Organic Farming in Ghaziabad, Uttar Pradesh in 2003, establishment of accreditation and certifying organization, financial support through different schemes, and projects, short-term and long-term training to organic farmers. To give more importance to organic farming it's included in postgraduate syllabus of agriculture by ICAR. M.S. Swaminathan, Chairman, Farmers Commission, India takes a great initiative by promoting organic farming policies. The Ministry of Commerce established National Organic Program in April, 2002 for exporting organic products. Government of India has taken several programmes or schemes:

(a) Rastriya Krishi Vikas Yojana

(b) National Mission for Sustainable Agriculture

(c) The Paramparagat Krishi Vikas Yojana

(d) Organic Value Chain Development in North Eastern Region Scheme

(e) The Mission for Integrated Development of Horticulture

(f) National Programme for Organic Production

(g) National Project on Organic Farming

The aforesaid projects/schemes, their aims, framework and the work profile are described below:

2.2.1 Rastriya Krishi Vikas Yojana

This scheme was launched in August 2007 as a part of the 11th five-year plan by the government of India. Launched under the aegis of the National Development Council, it seeks to achieve per cent annual growth in agriculture through development of agriculture and its allied sectors. The scheme is funded annually by a state plan scheme.

It was launched to promote organic farming and reduce dependence on chemical fertilizers. Under this scheme, soil health card was issued to 3 crore farmers in 2014 and 5.5 crore cards each during next two years. Additionally, 100 mobile soil testing laboratories are available in August 2014. Five new customized fertilizers have been approved, so that farmers get required nutrient for their fields.

Recently Indian government has launched a saffron park at pampering (Jammu and Kashmir). In this scheme facility for processing, packaging and the cultivation technology of saffron will be available (National Saffron Mission 2010-11).

2.2.2 National Mission for Sustainable Agriculture

Under this scheme the main features are:

(i) Established under departmental/ministerial coordination of government of India.

(ii) The main aim of this scheme to make organic farming sustainable, climate-resistant and ecofriendly and remunerative.

(iii) The scheme gives sufficient support to develop capacity building of organic farmers and stock holders.

(iv) IWMP, MGNREA and RKVY are inter-related schemes under its network.

(v) This will work towards the development of organic farming research and development sector.

(vi) This mission coordinates with soil health management.

This scheme is implemented by the state government Fertilizer Quality Control Board, National Center of Organic Farming and Soil, and Land Use Survey of India.

2.2.3 The Paramparagat Krishi Vikas Yojana

It is a cluster programme for the promotion of organic farming. Under this scheme 500,000 acres farming land is taken for a period of 3 years. More than 50 farmers make a cluster and farmers have no liability for expenditure on certification. According to the norms each farmer will get Rs. 20,000 per acre for a period of three years from seed sowing to harvesting of crops and extended up to bringing the crops in the market. The aforesaid scheme also extends its scopes up to domestic production and certification of organic products. Indian Government sanctioned Rs. 300 crore for the scheme in the year of 2015-16.

2.2.4 Organic Value Chain Development in North-Eastern Region Scheme

This scheme proposed a value chain based organic farming in north-eastern India with an investment of Rs. 115 crore. This mission has been launched with an aim to develop certified organic products with a value chain model which connects the growers and consumers. This scheme supports from the base, i.e., the financial inputs, seed certification, processing, brand building initiatives, marketing, etc. Approximately 30,000 to 50,000 farmers of north-eastern region are empowered through the establishment of 100 farmer producer organizations and organic product selling companies. In the year of 2015-16 against the budget allocation of Rs. 158.87 crore fund released was Rs. 112.11 crore. During 2016-17, against the budget allocation of Rs. 176.75 crore, fund released was Rs. 28.70 crore (Yadav, 2017). This organic value chain not only boosts organic entrepreneurship in the north-eastern region but also helps ecotourism across the country.

2.2.5 The Mission for Integrated Development of Horticulture

This plan was implemented in 2014-15 with an order from central government for the horticultural progress as per 12th plan. It is also associated with six sub-plans: National Horticulture Mission, National Bamboo Mission, Coconut Development Board, Central Institute for Horticulture, Nagaland, Horticultural Mission for North-East and Himalayan States, and National Horticulture Board. This mission provides technical support and state government regulatory support. It also encourages the aggregation activity of f arms like as FIAS/FPOS and FPCS. The main goal of this mission is to increase production, financial support, creation of jobs in the sector of organic agriculture. There are several accredited certification authorities. Among which

Indian Organic Certification Agency (INDOCERT) is one of the renowned organizations. INDOCERT with its associated partners in 2003 established International Competence Center for Organic Agriculture (ICCOA). ICCOA is responsible for various activities like increased organic networking, capacity building, market development and extension, awareness development, organize organic products trade fairs with other committee.

2.2.6 National Programme for Organic Production

Ministry of Commerce, Government of India proposed this programme in 2000 and implemented in 2001. Inspection and accreditation systems are its main objectives. Eleven states initially joined this scheme. The leading states are Maharashtra, Karnataka, Gujarat, Uttarakhand, Rajasthan, Madhya Pradesh, Nagaland and Mizoram, Tamil Nadu, and Sikkim. The programme provides information on organic production, standards, criteria, accreditation of inspection and certification bodies. Development of unique brand standard logo for organic products is also under its activity. From 2004, organic products of different crops were solely adopted and sold by various crop specific boards like Spices Board, Tea Board, Coffee Board, Coconut Board, Coca and Cashew Nut Board under the norms of government of India.

2.2.7 National Project on Organic Farming

Ministry of Agriculture, Government of India has proposed and launched a new scheme 'National project on Organic Farming' with an investment of Rs. 57.05 crore for the organic production. State governments implemented various agricultural policies to promote and encourage organic growers. Many state governments are introducing new organic farming policies. In India Maharashtra, Madhya Pradesh, West Bengal, Tamil Nadu, Kerala and Karnataka take an active role.

2.3 ROLE OF NGO AND OTHER ORGANIZATIONS TO PROMOTE ORGANIC FARMING

Non-government organizations play a crucial role to promote organic farming. A non-governmental organization (NGO) is any non-profit organization which is legally constituted, organized and operated on a local, national and international level and enhances the scope of organic farming. Classification of NGO is represented through Figure 2.1.

Six major activities of an NGO: protect environment, create awareness, protect human rights, encourage rehabilitation, combat manmade crisis, and gainful employment. In India approximately 53% NGOs are rural based and 47% are urban based. RUDSETI, SKDRDP,

MYRADA, AME, BAIF and KARUNA TRUST are some important NGOs. Major NGOs in the sector of agriculture are Nandi Foundation, A & A (action for social development), PRADAN, SRIJAN, ISAP, CARD, Aga Khan Foundation, DHAN FOUNDATION, GREENPEACE, Digital GREEN, AFC INDIA LIMITED, IFFCO FOUNDATION, GVT (Gramin Vikas Trust), OOHESION FOUNDATION TRUST, etc. ATMA Funds is also doing a great job to improve rural agriculture in training, job, funds, organic products supply, etc.

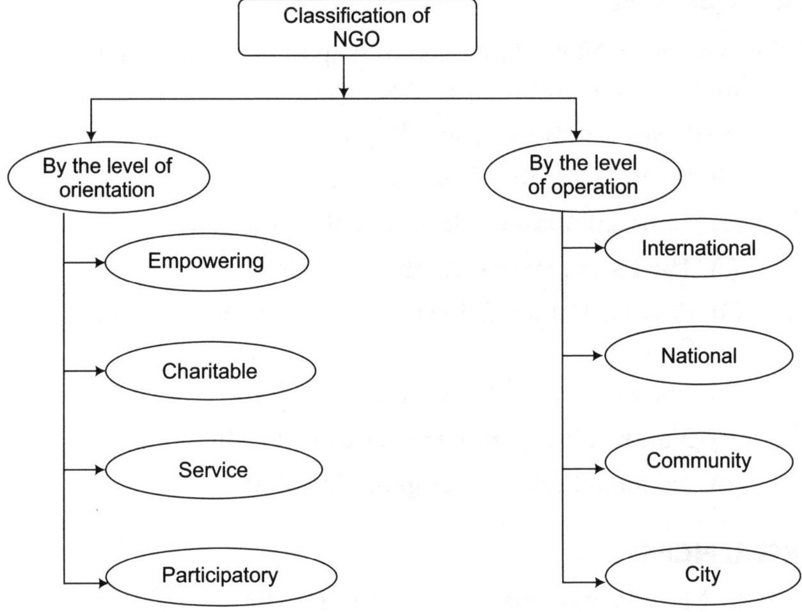

Figure 2.1 Classification based on orientation and operation of an NGO.

2.4 CONCLUSION

The emerging role of NGOs in rural development and for promotion of organic agriculture was described by Ramakrishna (2013). This intervention with high priority for organic agriculture, waste recycling, watershed management and value addition of agricultural/ horticultural products need to be promoted gains significance in the present scenario where we need to invigorate agricultural production and farmers livelihood in a sustainable manner (Surudhi and Jayakumar, 2016). This strategy may also help in organic agriculture. Various NGOs and self-help groups reach there where state/central government projects can't reach. Awareness camps, training, seed distribution, village fairs are the alternatives to promote organic agriculture. Overall allotment of revised budget, establishment of new laws for farmers' welfare and chemical pesticide restriction can bring sustainable faming and agricultural practices.

2.5 QUESTIONS

Long Questions

1. Briefly describe different government and non-government schemes to promote organic farming.
2. Briefly describe the role of NGOs in organic agriculture.
3. Write four government projects which help in organic farming.

Short Questions

1. What is an NGO? How does it help to promote organic farming in Indian agricultural perspective.

   Write short notes on the following:

   (a) Rastriya Krishi Vikas Yojana

   (b) National Mission for Sustainable Agriculture

   (c) The Paramparagat Krishi Vikas Yojana

   (d) Organic Value Chain Development in North-Eastern Region Scheme

   (e) The Mission for Integrated Development of Horticulture

   (f) National Programme for Organic Production

   (g) National Project on Organic Farming

REFERENCES

Surudhi, M. and Jayakumar, N. (2016). Role of local NGOs in promoting agricultural sustainability. *Indian Journal of Social Research:* 57(2): 217-222.

Ramakrishna, H. (2013). The emerging role of NGOs in rural development of India: An assessment. *International Journal of Social Science and Interdisciplinary Research,* 2(4): 43-51.

Yadav, M. (2017). Towards a healthier nation: organic farming and government policies in India. *International Journal of Advance Research and Development.* 2(5): 153-159.

3 Organic Ecosystems and Their Concepts

CHAPTER

3.1 INTRODUCTION

Organic farming is a natural farming system and is interlinked with various environmental components. In nature, there is no such conserve ecosystem which can tolerate interference with environmental components. Any kind of interference with the system does not affect only in one way but it necessarily evokes other often unpredictable reactions. Therefore, it's our responsibility to maintain our environmental components properly. Agro ecosystems are very much connected with more natural and near natural environment. Sometimes various negative impacts are found at juvenile or immature ecosystem, such as:

- It always supports by providing additional energy inputs.
- Sometimes reduced biodiversity
- In many cases it provides artificial support to dominant species
- Moderate anthropogenic disclimax or juvenile stage of succession.
- Makes barrier of self-regulatory processes
- Environmental stability decreases
- Occurrence of irreversible degradation process

The external moderation of the associated agro-ecosystem is a high degree of landscape urbanization (natural vegetation suppression, the amount of built-up areas and sharp land boundaries). At field level it is characterized with stand uniformity, inability to self regulation, often poor environmental adaptation, permanent soil erosion and the need to control other material and energy inputs. Intensive agro-ecosystems showed spatial landscape heterogeneity reduction and the corresponding species reduction. Combining activity can generate intensification. Creation is barely particular (for number of developed plant species is diminishing to monocultures, with animals, the specialization goes down to level of individual classification rearing without any connections to the land and harvest creation). Various intensification factors are - step land use (multistorey stables), concentration (production organism's density increase in time and space), also improved mechanistic to technical approaches and intensive use of additional chemical inputs, energy and information. Exceptionally escalated motorized framework is getting totally reliant on outside sources of info (apparatus, fuel, synthetic compounds, and seeds). High outer energy material information sources firmly diminish the frameworks' energy effectiveness. The proportion of energy contribution to energy picked up from the harvest is up to 3:1 while with non-escalated frameworks, it is 1:20 and that's just the beginning. Inside exceptionally concentrated motorized domesticated animals creation framework, the energy balance is even less successful. In any case, these frameworks are exceptionally powerful in the present moment regarding work efficiency and land use.

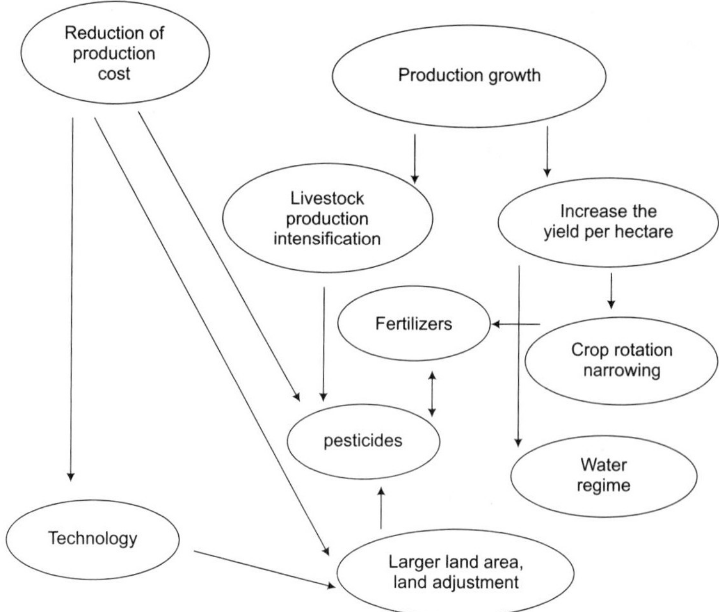

Figure 3.1 Schematic representation of relationship between market based production and agro-ecosystem.

The terminology organic ecosystem clearly explains, plainly the feasible agroecosystem doesn't convey just the capacity of food and column material creation. The natural cultivating framework is centred around the homogenization of scene creation and non-creation capacities, where the accentuation is laid on ecological perspectives. Schematic representation of relationship between market based production and agro-ecosystem is shown Figure 3.1. Remember for specific:

1. Upkeep and improvement of soil fertility.
2. Supplements distribution and an avoidance of the section of incidental substance into agroecosystem.
3. Save the water balance and keep it pollution free.
4. Air quality improvement and ozone harming substance discharge decrease.
5. Hereditary assets insurance and biodiversity upkeep.
6. Protection of scene highlights and their harmonization.
7. Proficient utilization of energy, zeroing in on sustainable assets.
8. Streamlining of life for all creatures, including people.

Organic farming depends on the standards of feasible cultivation and hence it is a model for the manageable agro-ecosystem's foundation.

3.2 SOIL ECOLOGY AND ENVIRONMENTAL IMPACT

Soil is one of the most crucial resources and plays an important role in agriculture. Healthy soil is essential for better plant growth and nutrition. Additionally, soil has multidimensional functions like buffering, filtering and transformation and aside it's a hub of beneficial microorganisms. Some of the positive changes in soil are mentioned below:

(i) More than 30% organic carbon content in soil organic matter.
(ii) By using organic amendment soil biological activity can increase by 30-100%, which is a biomass decomposition indicator.
(iii) Edaphon biomass can increase up to 50-80%.
(iv) Higher saprophytic fungi colonized at the roots of higher plants and makes mycorrhizal association.
(v) Soil microorganisms use soil organic components more efficiently.
(vi) Soil physio-chemical property and soil structure, porosity will improve.
(vii) Improved hydroscopicity and reduction in soil erosion level.

Research on soil organic matter concentrates on the organic carbon content and its changes during conversion to organic farming. Several studies have confirmed that areas under organic cultivation have a higher amount of organic carbon content as compared to areas under conventional cultivation. However, in some researches, there was a higher decomposition of organic matter. More intensive soil cultivation helps in mechanical weed control. However, long-term experiments have confirmed the hypothesis that organic farming methods better protect the soil organic matter. The research also points to a larger amount of humic substances. Soil organic matter protection can only be possible through soil cultivation. In accurately designed crop structure, fertilizers have the equal importance. It is found that structural soil is more resistant to erosion (Siegrist et al., 1998).

Higher supply of organic matter and organic fertilizers create favourable living conditions for soil fauna. The soil provides a habitat for a large number of microorganisms. Crucial role of microorganisms is decomposition of organic matter and transformation of inorganic substances where nutrients are more accessible for plants along with synthesis of complex organic substances enriching humus reserves in the soil biota. In organic farming, the main role for nitrogenous nutrients comes from papilionaceous plants. In case of nonsymbiotic fixation, e.g., by the soil-free living heterotrophic aerobic bacteria. Rhizosphere is a zone where the main part of nutrient cycle takes part due to the interaction between soil, roots and microorganisms colonizing the plant root environment. In organically cultivated soils, more than 40% mycorrhizal association is found (Mader et al., 2002).

3.3 IMPACT ON NUTRIENT RECIRCULATION

In agricultural ecosystems soil fertility increases by additional inputs such as manure or fertilizer. Plant nutrition within conventional farming is more dependent on the input of nutrients in the slightly soluble form, predominantly from synthetic fertilizers. A part of nutrients leaves the system as a loss. On the contrary, the natural ecosystem fertility depends almost entirely on natural biological processes like nitrogen fixation and soil organic matter mineralization. Organic farming stands on the concept of the living system. Therefore, the fertilization system is designed to respect natural nutrient cycle and not to adversely affect complex biological processes which nutrient cycles are dependent on. Organic approach is the continuous source of organic matter content in the soil while conventional intensive farming is not capable to mitigate the same matter content. Leguminous and clover crops by successive crop rotation serve soil nitrogen to a considerably because of their agromeliorative effect on the soil. The results is found in physical properties with an effect on the soil bulk density, water holding capacity, soil structure stability, increasing porosity, etc. The size of invertebrate organism population depends on soil physio-chemical properties with well aerated non-compacted soil with low density.

Organic farming policies encircled with the most closed nutrient cycle, provide minimal nutrient loss and limited nutrient supply to the system. For maintaining of the soil production capacity, it is important to replace nutrients drawn from soil by harvests and lost nutrients by biologically transformed organic matter in the soil. Regular supply of organic matter into the soil helps to enrich soil's organic matter (clover, catch crops for green manure), crop residues, residues of cultivated plant roots and manure. Soil organic matter serves as a continuous reservoir of nutrients and energy for the soil environment. This is also considered for soil environment stabilization. Soil organic matter is an important source of nutrients for grown plants, source of energy for soil microorganisms, improves physio-chemical soil properties, water regime, increases decontamination and buffering soil capacity and decreases nutrient losses washed away from soil, increases antiphytopathogenic soil potential and strengthens the plant immune system.

3.4 WATER IN LANDSCAPE EFFECT

Disposal of pesticides and morfo-regulators from the organic farming system reduces contamination of the environment. Alternately, in the area of organically cultivated crop fields, surface water and groundwater are less contaminated with plant protection products. These substances also harm the aquatic animals, even at low concentrations (Liess et al., 2001). Synthetic nitrogen fertilizer reduces significantly the load of surface and groundwater by nitrates. Approximately 35-64% nitrates are washed away through organic farming as compared to conventional farming practices. There were two main areas of water pollution in organic farming, manure composting and farming with residual nitrogen from leguminous plants. Contamination of groundwater and surface water occurs through storage and composting of manure with subsequent leakage. Nutrients from intensively cultivated cropland load water due to overland flow. Within organic farming, this risk is reduced by greater ruggedness of landscape, more extensive integration of landscape features as well as the crop diversity and their optimal distributing. Marginal strips along water courses and reservoirs and protective grass strips on slopes make buffering areas, limit overland flow and prevent erosion and expand biodiversity. Organic farming is supposed to be the preferred farming system especially in the areas of water resources conservation. Organic farmers fertilize the soil in such way as not to pollute groundwater. In organic farming in addition to manure and liquid manure, green manure is also used as fertilizer and legumes are properly incorporated into the crop rotation. This reduces leaching of nitrogen into groundwater. As a result of leaching through the soil profile and due to erosion and surface runoff, nitrates in water cause contamination of the hydrosphere and along with phosphorus cause eutrophication. Concurrently, there is a leaching of base cations

(K+, Ca2+) and thus the upper soil layers are depleted of these nutrients. Indirectly, this leads to acidification. Due to leaching, several tonnes of N is lost. Organically cultivated areas provide better flood protection than conventional surfaces. The high infiltration capacity of the soil with virgin structure may reduce the intensity of floods.

3.5 IMPACT ON AIR QUALITY

Agricultural activities have a large effect on air quality. Organic farming as a whole contributes to the creation of anthropogenic greenhouse gases with about 14% while the ratio differentiates with particular countries according to the agricultural production intensity. Due to its large area impact, agriculture belongs to the largest producers after industry and mining Cole et al. (1997) state that agriculture contributes to annual increase of GHG emissions with approximately one fifth. Even higher value is stated by Cerri et al. (2009), whose findings report the proportion of 27%.However, with the increasing consumption of food and agricultural intensification, this percentage is rising. When adding pre-farming and post-farming phases to the agricultural frame itself or quantifying food life cycle, the emission load is even higher. This is mainly due to production of agrochemicals and processing of primary agricultural production. Moreover, due to the increasing human population, agriculture will even increase its pressure on the environment. There is constantly running the conversion of natural habitats into agroecosystems and in parallel the intensification of farming on the existing agricultural land. This is largely accompanied by other chemicalization of agriculture. The pressure on increasing yields and the food consumption grows share of agricultural emission load production. However, in most cases, organic farming produces lower emission load, not only in the field phase but also in the consequential phases. Agriculture produces emissions in many ways. For example, CO_2 is released during the consumption of fossil fuels or in reduction of organic matter content in the soil. N_2O is released as a result of fertilizer application and in soil processes, CH_4 from the digestive tract of some livestock species. Especially in the crop production, the emission is influenced by the intensity and thus by the system of farming.

3.6 EFFECT ON BIODIVERSITY

The positive impact of organic farming on biodiversity is based on an effort to extend the range of cultivated crops and livestock and thereby to increase a genetic, species and ecosystem diversity. On this basis and on the basis of environment-friendly agroecosystem management, the functional agrobiodiversity has increased. Growing biodiversity at all levels (predators, parasites, wild plants, pollinators, soil fauna and flora) supports the ecosystem functions (population control, competition, allelopathy, organic matter decomposition, nutrient sorption and their cycles). It contributes to the agroecosystem's stability and sustainability. It improves resistance of production organisms against harmful agents gradation and contributes to their effective control, improves the nutrient utilization in agroecosystem and reduces eutrophication. It contributes to the erosion reduction and improves the moisture use, helps to increase diversity and abundance of wild flora and fauna in the landscape. In the organically cultivated farms, there are more than 85% greater number of plant species, a third more bats and there live about 17% more spiders and 5% more bird species (Fuller et al., 2005; Hole et al., 2005). On the organic field, there are, for example, nine times more species of plants and accompanying weeds growing, there live 15% more ground beetles and 25% more earthworms than in the fields with integrated farming. Greater variety of plants, hedges, grassy field margins, smaller areas of land, smaller corn ear density, area gardening, stubble and green soil cover in winter create favourable conditions, e.g., skylark. Already after one year of transitional period, the number of skylark nests has doubled. Nesting swallows and birds of prey also give priority to food from organic areas. In autumn and in winter, there were significantly more seeds and insects for songbirds and also more food for birds of prey in organically cultivated fields.

3.7 CONCLUSION

Predominantly, the farming systems have still been considered production systems. The typical conventional agroecosystems are characterized as very open, as systems based on technological processes substituting the ecological ones, systems with high labour productivity but with lower biodiversity, flexibility, stability and sustainability on the other hand. On the contrary, the organic farming systems are based on sustainable development principles and the holistic world approach. It is a production system focused on preservation and improvement of natural resources and the environment at the same time. As for the system concept, there is an effort to balance the economic, environmental and social aspects and relations on global and local level. Agricultural activity itself is considered a process of reasonable ecosystem exploitation with respect to its stability and sustainability. Just this divergence between the both approaches, the intensive industrial agriculture on one hand and the sustainable organic farming systems on the other, makes them very different from the environment point of view. In the organic farming systems the soil quality remains the main interest. Soil erosion control measures include cover crops, mulching, limited soil cultivation, windbreaks planting, use of lighter and smaller machinery, keep an optimal soil structure and looseness. These soil cultivation principles are combined with lower need of pesticides. Thus, there is a sufficient nutrient cycle and organic matter content in the soil leading to an optimal soil's biological activity and fertility. Generally, the principles of organic farming ensure protection of water sources and soil moisture, prevent the underground and surface waters from pollutants and sediments as well. Water preservation is the priority, there are used terraces, environmental corridors and border zones and other measures. There are also considerable differences in biodiversity, which, in the conventional farming, markedly suffer not only due to crop range reduction leading to monocultures in fact, but also thanks to reduction of associated fauna and flora thought as harmful and thus systematically eliminated or suffering from pesticides or other biocides at the same time. By contrast, organic farming purposefully supports biodiversity, takes advantage of more adaptable animal and crop species and varieties to the habitat conditions, uses varied crop rotations, species and variety of mixtures, applies technical and organizational measures friendly to the organisms and the environment. Organic farming systems are also more environment-friendly with regard to the greenhouse gas emissions. Above all, this emission reduction is achieved thanks to the limited use of synthetic fertilizers and pesticides and lower livestock production intensity.

3.8 QUESTIONS

Long Questions

1. Briefly describe the concept of organic ecosystem.
2. How soil ecology and environment interlinked with organic ecosystem?
3. Describe any two environmental components interlinked with organic ecosystem.

Short Questions

1. Compare the rate of greenhouse gas emissions in traditional farming system and organic farming system and which system is more effective.
2. How water, soil and air quality is interlinked with organic ecosystem.
3. Describe organic ecosystem on impact on biodiversity.

REFERENCES

Cerri CC., Maia SMF., Galdos MV., Cerri CEP., Feigl BJ., Bernoux M. 2009. Brazilian greenhouse gas emissions: the importance of agriculture and livestock. *Scientia Agricola,* 6: 831-843.

Cole CV., Duxbury J., Freney J., Heinemeyer O., Minami K., Mosier A. et al., 1997. Global estimates of potential mitigation of greenhouse gas emissions by agriculture. *Nutrient Cycling in Agroecosystems,* 49: 221-228.

Fuller RJ., Norton LR., Feber RE., Johnson PJ., Chamberlain DE., Joys AC. et al. 2005. Benefits of organic farming to biodiversity vary among taxa. *Biology Letters,* 1(4): 431-434.

Hole DG., Perkins AJ., Wilson JD., Alexander IH., Grice PV., Evans AD. 2005. Does organic farming benefit biodiversity? *Biological Conservation,* 122: 113-130.

Liess M., Schulz R., Berenzen N., Nanko-Drees J., Wogram J. 2001. Pflanzenschutzmittel-Belastung und Lebensgemeinschaften in Fliessgewassern mit landwirtschaftlich genutztem Umland. Berlin: Technische Univerität Braunschweig.

Mader P., Fliessbach A., Dubois D., Gunst L., Fried PM., Niggli U. 2002. Soil fertility and biodiversity in organic farming. *Science,* 296: 1694-1697.

Siegrist S., Schaub D., Pfiffner L., Mader P. 1998. Does organic agriculture reduce soil erodibility? The results of a long-term field study on loess in Switzerland. Agriculture, *Ecosystems and Environment,* 69: 253-265.

Unit 2

Nutritional Entity of Organic Farming

CHAPTER 4 Organic Nutrient Resources and Their Fortification

4.1 INTRODUCTION

Organic nutrient sources like manure, green manures, compost, meals and so forth—recommendations for nutrient needs is highly demanding for plants. Soil testing only indicates the deficiency or pollutant levels of soil, not to recommend any organic amendments in the soil. This is due to the percentage available components/nutrients in such materials is highly variable and usually depends on its source, storage methods, and the amount and type of materials used to make the nutrient source. Low nutrient content and often limited sources of many organic components create difficulty in developing a specific recommendation.

As an essential contribution to the present-day agribusiness, manures make a significant commitment to the achievement of high harvest yields (Erisman et al., 2008). Agricultural efficiency expanded in a large number of areas that accomplished the Green Revolution through the utilization of nitrogen, phosphorus, and potassium (NPK) containing manures. Although, the reducing aspect of productivity under less favourable biophysical conditions and nutritional deficiencies resulting from the consumption of crop produce with low dietary values directed the search towards more sustainable and appropriate solutions to produce crops with higher yields and nutritional quality. Together with NPK, calcium (Ca), magnesium (Mg), and sulphur (S) are classified as essential macronutrients. The eight other essential nutrients constitute a distinct group of elements required by plants in very small amounts, described conventionally as micronutrients: namely (copper [Cu], iron [Fe], manganese [Mn], molybdenum [Mo], nickel [Ni], zinc [Zn]), boron [B], and chlorine [Cl]). Still, other elements like selenium (Se), silicon (Si), and sodium (Na) are regarded as nonessential, although they have been found to enhance growth and confer other benefits to plants (Marschner 2012). Nutrients also play an important role in humans and, in many cases, dictate our psychological development and how we respond to diseases, especially the micronutrients. In fact, many of the enzymes, proteins, and other biological compounds that perform important metabolic functions in humans cannot accomplish such functions without micronutrients. Different countries where staple foods consist mainly of cereals, roots, and tubers grown in nutrient-poor soils, human micronutrient deficiency is widespread. Although, micronutrient amendments in soil for crop uptake can lower the impact of their deficiency in humans (Cakmak 2008), much like the case of Se in Finland, where fertilization of soils with Se-fortified NPK not only improved food crop yields but also improved daily Se intake and serum Se concentrations in the Finnish population (Ekholm et al., 2005).

4.2 NATIONAL ORGANIC STANDARD SUMMARY ON SOIL FERTILITY MANAGEMENT

The goals of all classes of growers are to maintain soil fertility and to provide sufficient nutrients to the crops grown, maintain or improve the soil condition, and minimize erosion. Fertilizers and soil amendments which are in compliance with the national organic standards are available to complement other fertility practices. Additionally, mined substances of low solubility can be used to supply plant nutrients. Plant or animal ashes can also be used to improve soil fertility as long as they have not been combined or treated with a prohibited substance and are not themselves a prohibited substance.

4.3 MICRONUTRIENT DEFICIENCIES IN SOIL

Soil contains significant amounts of micronutrients, their availability to plants could be maintained, under the influence of different edaphic and biological factors such as pH, competing cations, anions, organic matter, soil geomorphology, soil parent materials, and soil microbiology, etc. Some micronutrients when come in contact with soil undergo rapid reaction with compounds such as phosphates and carbonates, form chemical precipitates, or they may interact with clay colloids and other mineral complexes, rendering them unavailable to the crop (Allen 2002; Marschner 2012). Furthermore, similarities in the biological uptake mechanisms of micronutrients by plant roots also influence the ultimate plant bioavailability of specific micronutrients. Nevertheless, soil micronutrient availability could be modulated by changing the soil pH, such as through liming with dolomite. Also, specific plants may scavenge the soil for micronutrients through adaptation strategies such as elaboration of extensive root systems, and importantly, the excretion of exudates, including organic acids and phytosiderophores, to dissolve fixed minerals (White et al., 2013; Keuskamp et al., 2015). One common role of micronutrients in different global agroecosystems is the fact that their deficiency in crops is compounded by a low crop use efficiency (i.e., low crop response per unit of micronutrient, relative to no micronutrient application), typically <10 %, compared to between 20 and 80%for N, P, and K (Baligar et al., 2001). In Africa and Asia, it is reported that majority of the arable land is affected by multiple micronutrient deficiencies (Oliver and Gregory 2015). In these regions, the scenario is further complicated by the inadequate replenishment through fertilization of micronutrients biologically mined from the soil by plant roots, which has had negative agronomic consequences for crop productivity. For example, the use of NPK fertilizers in India for growing rice and wheat over many decades contributed to high crop yields but also resulted in crops extracting micronutrients from the soil to the extent that the bioavailability of micronutrients such as Zn has become so limited that it poses problems for Zn nutrition of humans (Monreal et al., 2015). Not surprisingly, the application of micronutrients has been found to ameliorate micronutrient deficiency and, consequently, improve crop yield levels.

4.4 SOIL TESTING

Good quality crop yields can be expected over a rather wide range of nutrient levels in most soils. The most desirable level of each nutrient depends on such variables as the amount of rainfall, temperatures, amount of sunlight, soil texture, soil drainage, prevalence and severity of plant diseases, and the crop cultivar grown. The more intensive the type of production, the more important information is provided by a soil test.

4.5 SOIL TEST VALUES

Soil test values are interpreted as 'deficient', 'optimum', or 'exceeds crop needs'. Crops normally produce best when nutrients are present in balanced amounts at 'optimum' levels. Application of a nutrient shown to be 'deficient' should increase yield. When levels fall in the 'exceeds crop needs' category, more than enough of a nutrient is present.

4.6 WHEN NUTRIENT LEVELS EXCEED CROP NEEDS

Soil nutrient levels that exceed crop needs can be as bad as deficient levels. High soil nutrient levels might not only represent an economic loss but may also result in crop, animal, or environmental problems. To achieve maximum yield and quality from each crop, a specific balance among elements is necessary. Very high phosphorus levels (above about 310 pounds P_2O_5 per acre or 140 pounds of phosphorus per acre for vegetable crops) in the soil may lead to deficiencies of other nutrients, especially of iron and zinc. If potassium, magnesium, or calcium is high, serious nutrient imbalances can occur. When potassium levels are above about 5 per cent saturation, magnesium levels 15 per cent, and calcium levels 80 per cent, soil nutrition is begin to get out of the optimum range. Use best management practices to avoid increasing nutrient levels that already exceed crop needs. Yield and quality are likely to be reduced by reapplying a nutrient already present in very high amounts.

4.7 BALANCE AND IMBALANCE OF NUTRIENTS IN ORGANIC NUTRIENT SOURCES

An unbalanced plant-nutrient status can be the result of using either organic or inorganic fertilizers. Most organic materials (including compost) do not contain nutrients in balanced amounts as needed by plants. In particular, many nutrients from animal sources (e.g., manure) have an excess of phosphorus and potassium relative to plant demand for nitrogen. In soil these nutrients can accumulate to levels exceeding crop needs with repeated application based on plant nitrogen needs. When using organic materials, test soil regularly to monitor phosphorus, potassium, and salt accumulation. Nutrient amendments should also be tested regularly, as similar organic materials may vary considerably in nutrient content depending on their source, handling, and conditions present when the plant or organism was living. The use of other sources of plant nutrients may be necessary to correct imbalances (for example, legume green manure crops that contribute nitrogen without increasing phosphorus and potassium).

4.8 NUTRIENT AVAILABILITY FROM ORGANIC NUTRIENT SOURCES

Suggested amounts of organic amendments or fertilizers to be applied in lieu of chemicals or inorganic fertilizers may or may not be equally effective because of differences in the physical and chemical nature of organic materials. Most nutrients from organic materials are less available to plants compared to inorganic fertilizers.

4.9 INCREASING SOIL PH, CALCIUM AND MAGNESIUM LEVELS

If soil test analysis indicates a need for a liming material to increase soil pH, calcium, or magnesium levels, the use of these materials is the first step in soil management. Whenever possible, a liming material should be applied during fall prior to planting to provide several months for these materials to begin reacting with soil particles. The liming potential of wood ashes (32 per cent calcium oxide, CaO) is such that 3 pounds of wood ash is equal to 2 pounds of ground limestone (50 per cent CaO). If soil pH is 6.8 or above, wood ashes should not be used.

4.10 DECREASING SOIL pH

Soil test analysis reports if indicate that the soil pH exceeds optimum, it will need to be lowered. Some materials for lowering soil pH include sulphur, peat moss, and cottonseed meal. For large areas sulphur is likely the most economical option. Elemental sulphur and iron sulphate can be used in organic production. Both of these products are labelled "restricted", which means that they are allowed in organic production but are subject to meeting certain criteria. Sulphur reacts slowly with soil particles; therefore, allow several months for changes in soil pH to occur. Whenever possible, apply sulphur well in advance of planting to provide sufficient time for reacting with soil particles.

4.11 RECOMMENDATION FOR NITROZEN, PHOSPHATE AND POTASH

Fertilizer recommendations for crop production are expressed in amounts of N (nitrogen), P_2O_5 (phosphate), and K_2O (potash) per acre. A soil with a deficient phosphorus level may present a problem when relying exclusively on organic materials. Rock phosphate is the most economical organic source of phosphorus allowed in organic farming. Raw rock phosphate, even when it is from high-grade minerals and finely ground, has been quite ineffective on soils with a pH much higher than 6.0, due to low solubility and exceedingly slow reaction time. Rock phosphate needs acids to bring about the release of P_2O_5 (phosphate) for plant use; hence, it is most efficiently used on acid soils. In acid soils P_2O_5 may be available to plants in subsequent seasons following application. Although rock phosphate contains 20 to 32 per cent P_2O_5, available P_2O_5 is only 5 per cent. A number of organic materials may be used to supply N and K_2O. Before using any listed materials, it is best for organic growers to check with a certifying agency to ensure that organic certification is not compromised by their use.

4.12 CALCULATING SODIUM NITRATE (CHILEAN NITRATE) THAT CAN BE APPLIED ON ORGANIC FARMS

Sodium nitrate, also known as chilean nitrate, cannot account for more than 20 per cent of the N requirements of organic crops. Its use is also prohibited by the International Federation of Organic Agriculture Movements (IFOAM). To calculate the amount of sodium nitrate, first determine the amount of nitrogen recommended for the crop. This information varies by crop and can be found on soil test reports. Next, multiply the recommended rate by 0.20 (20 per cent) to determine how much of the recommended nitrogen can be satisfied by sodium nitrate.

4.13 ADDING NITROGEN WITH LEGUMINOUS GREEN MANURES

Green manures are crops that are turned into the soil while they are young and succulent, rather than harvested. They can be grown for different purposes: adding nitrogen to the soil, suppressing weeds, scavenging nutrients left in the soil, or increasing the soil organic matter content. Legume crops can add nitrogen to the soil. They are able to establish relationships with soil-borne bacteria that are capable of extracting nitrogen gas from the atmosphere and converting it into a form that the plant can use. Seed of legumes may need to be inoculated with these bacteria; inoculants are commercially available. Inoculants are specific to the legume species grown, so choose a compatible inoculant. Also, not all brands of inoculants are allowed in organic production. Organic growers should check with their certifying agency before using it.

4.14 SOIL ORGANIC MATTER CONTENT

Most crops grow best in soils with organic matter contents between 2 and 5 per cent. Organic matter serves beneficial functions, including minimizing soil temperature fluctuations, serving as a nutrient warehouse, buffering the soil to changing pH, and increasing the ability of the soil to hold nutrients. Additionally, soil structure can be improved along with the ability of the soil to hold water and air. Organic matter can also provide habitat for beneficial soil microorganisms. A soil with an optimal organic matter content is better able to tolerate adverse conditions. For example, during drought conditions or when excess water is present, a soil with a good organic matter content will rebound quicker than one with a low organic matter content.

4.15 MINERALIZATION

Nitrogen for plant use is released from organic matter as it is broken down by soil microorganisms, a process formally called mineralization. The rate of mineralization is influenced by many factors including environmental conditions (soil temperature, soil moisture, light levels, etc.), tillage practices (soil incorporation, depth of incorporation, timing of tillage, etc.), soil microorganism populations, and composition and particle size of the organic matter. To complicate matters, these factors also interact. In general, mineralization will be slower at lower soil temperatures because the soil organisms involved are less active at low temperatures. Mineralization is slow when soil is dry or waterlogged for the same reason. When organic nutrient sources, such as composts and manures, are not turned into soil and left on the soil surface, mineralization will be relatively slow due to drying. When the organic nutrient source is incorporated into the top 6 to 8 inches of soil, mineralization will occur more rapidly because this is the area of soil where most of the organisms involved in mineralization are. When the organic nutrient source is soil incorporated deeper than 8 inches, mineralization will occur more slowly because lower oxygen levels at deeper soil depths limit the number of organisms involved in mineralization.

4.16 USING COMPOST

Composts can be an important part of managing nutrients. In addition to adding nutrients to the soil, they can improve long-term soil quality. Composts are best when used in combination with other nutrient management strategies including green manures, other fertilizers, and crop rotations.

4.17 NATIONAL ORGANIC STANDARD SUMMARY ON COMPOST

Composted plant and animal materials can be incorporated into soil as necessary, provided the compost meets carbon to nitrogen (C:N) and temperature requirements. Compost used must have had an initial C:N ratio between 25:1 and 40:1. In addition, when using an in-vessel or static aerated pile system for composting, the pile must reach a temperature between 131°F and 170°F for a minimum of 3 days. If using a windrow system for generating compost, the pile temperature must be maintained between 131°F and 170°F for a minimum of 15 days and turned a minimum of five times during that time.

4.18 NITROZEN, PHOSPHORUS AND POTASSIUM IN COMPOST

Nitrogen in compost is in two forms: the organic form and the inorganic form made up of ammonium and nitrate. Total nitrogen content can be determined by adding the organic and inorganic forms of nitrogen (organic nitrogen + ammonium nitrogen + nitrate nitrogen = total nitrogen). Total nitrogen will normally range from 0.5 to 2.5 per cent (dry weight basis) in finished composts. The ammonium and nitrate forms are readily available for plant uptake. However, the organic form of nitrogen must be broken down through the process of mineralization into inorganic nitrogen for plant use. The rate of nitrogen mineralization depends on many factors (including, pH, depth of incorporation, compost quality and quantity, climate conditions, particle size, level of decomposition, soil microorganisms, C:N, etc.) and varies widely.

A typical compost might have an analysis of 1-0.7-1. Using 12.5 tonnes per acre of this compost applies 250 pounds of nitrogen, 175 pounds of phosphate, and 250 pounds of potassium. Additionally, magnesium and calcium will be added to the soil. Nutrient levels can quickly surpass optimum levels when using compost yearly. This is a problem for several reasons. Yields can be below optimum as a result of nutrient imbalances in the soil. Research has also shown that different weeds proliferate when certain nutrients are available in excess. Nitrogen and phosphorus buildup can also be an environmental hazard.

4.19 CARBON TO NITROZEN RATIO (C : N)

The carbon to nitrogen ratio (C:N) may be used as an indicator of compost stability and nitrogen availability. Compost C:N ratios typically decrease during composting if the starting C:N ratio is >25 but may increase if starting C:N ratios are low (<15) and nitrogen is lost during the composting process. Composts with high C:N ratios (>30) will likely tie up (immobilize) nitrogen if applied to soil, while those with low C:N ratios (<20) will mineralize organic nitrogen into inorganic (plant-available) nitrogen.

4.20 TIPS FOR USING COMPOST

1. Avoid the continuous use of compost or any single organic nutrient source containing more than one nutrient. Instead, use a variety of nutrient sources. This will help avoid reaching soil nutrient levels exceeding crop needs.
2. Use soil testing to keep track of soil nutrient levels. If levels exceed crop needs, avoid using compost. Instead, use a nutrient source that has no or minimal levels of the nutrient(s) in excess. One exception to this tip is that starter phosphorus may be needed for some crops when soils are cold in early spring, even when soil phosphorus levels exceed crop needs.
3. Use compost testing. Composts differ in their chemical analysis. By having it tested, it can be more accurately applied.
4. Calculate the amount of compost to apply. As noted in the "Nitrogen, Phosphorus, and Potassium in Compost" section, this is commonly based on the nitrogen needs of the crop unless phosphorus levels exceed crop needs. When phosphorus

levels are high, calculate both phosphorus and nitrogen based application rates and use the lower of the two. This practice will help avoid over application and its associated costs. The cost of the compost, environmental costs, and loss of profits due to compromised plant health.

5. Incorporate compost in the soil. This will promote the mineralization process and minimize runoff and erosion losses.
6. If phosphorus levels exceed crop needs or it is suspected that nitrogen levels are high, minimize losses through erosion, runoff, and leaching. Environmental concerns develop when phosphorus and nitrogen reach bodies of water. Minimizing erosion by planting cover crops, using reduced tillage practices, or using grass waterways to catch and infiltrate runoff from a field can curtail movement of these nutrients.

4.21 ESTIMATING HOW MUCH COMPOST TO APPLY BASED ON CROP NITROZEN NEEDS

Two basic methods for estimating compost application rates exist. Both methods require knowing the nitrogen content (the organic and inorganic nitrogen) of the compost. Regional soil testing lab gives the report of the sampled soil and the need to add organic or inorganic form of nitrogen.

4.22 CONCLUSION

Micronutrients are very important for plant growth. Although, micronutrient deficiency is considered to be a big problem for sustainable agriculture and organic farming. Micronutients are so far used as fertilizer to increase productivity. Especially when the application of conventional NPK fertilizers is not efficient. Gradual reduction of micronutrients level decreases soil fertility. To maintain soil fertility there needs to be a periodical source of various micronutrients. A single fertilizer will not fulfil all the micronutrient deficiency. Various combinations of micronutrient products can develop a new fertilizer. But always remain careful before soil treatment because soil analysis and plant specific micronutrient deficiency is important. By thinking about all classes of farmers, the product should be affordable to purchase. Micronutrients have an immense role in crop biomass production, enhancing yield, suppressing crop diseases and fortifying crops. Development of soil health card is a good initiative from Indian Government. Various soil testing laboratory and agencies help under this government project. Farmers education programme based on plant diseases and disorders linked with micronutrient deficiency is highly appreciating.

4.23 QUESTIONS

Long Questions

1. Briefly describe organic nutrient resources and fortification.
2. Briefly describe various nutritional deficiencies.
3. Discuss briefly government rules and regulations on fertilizer doses.

Short Questions

1. Where do micronutrients come from?
2. How to maintain balance and imbalance of nutrients in organic nutrient sources?
3. How to increase soil pH, calcium and magnesium levels?

REFERENCES

Baligar VC., Fageria NK, He ZL. 2001. Nutrient use efficiency in plants. *Commun Soil Sci Plant Anal,* 32: 921–950. doi:10.1081/CSS-100104098

Cakmak I. 2008. Enrichment of cereal grains with zinc: agronomic or genetic biofortification? *Plant Soil,* 302: 1–17. doi:10.1007/s11104-007-9466-3

Erisman JW., Sutton MA., Galloway JN., Klimont Z., Winiwarter W. 2008 How a century of ammonia synthesis changed the world. *Nat Geosci,* 1: 636–639. doi:10.1038/ngeo325

Ekholm P., Eurola M., Venalainen E-R. 2005 Selenium content of foods and diets in Finland. In: Eurola M (ed) Proceedings twenty years of selenium fertilization. *Agrifood Res Rep,* 69: 39-45

Keuskamp DH., Kimber R., Bindraban PS., Dimkpa CO., Schenkeveld WDC. 2015 Plant exudates for nutrient uptake. *VFRC Report* 2015/4. Virtual Fertilizer Research Center, Washington DC, USA, pp. 53

Marschner P. 2012 Marschner's Mineral Nutrition of Higher Plants, 3rd edn. Elsevier, Oxford.

Monreal CM., DeRosa M., Mallubhotla SC., Bindraban PS., Dimkpa CO. 2015 Nanotechnologies for increasing the crop use efficiency of fertilizer-micronutrients. *Biol Fert Soils,* doi:10.1007/s00374-015-1073-5

Oliver MA., Gregory PJ. 2015 Soil, food security and human health: a review. *Euro J Soil Sci,* 66:257–276. doi:10.1111/ejss.12216

White PJ., George TS., Dupuy LX., Karley AJ., Valentine TA., Wiesel L., Wishart J. 2013 Root traits for infertile soils. *Front Plant Sci,* 4: 193. doi:10.3389/fpls.2013.00193

5
CHAPTER

Restrictions to Nutrient Use in Organic Farming

5.1 INTRODUCTION

In organic farming system nutrient management is an important aspect. Nitrates (processed soluble fertilizer) which feed the plant directly and are thought to bypass the natural processes of the soil, aren't generally acceptable. Nutrient supply to crop plants is supported through recycling, the management of biologically-related processes like organic process by clover and other legumes, and therefore the limited use of unrefined, slowly-soluble off-farm materials that decompose within the same way as soil minerals or organic matter. The aim is to achieve as far as possible a closed nutrient cycle of the farm and to minimize adverse environmental impact. Effective management of any 'waste' materials such as manures and crop residues may be a key to nutrient cycling on organic farms. However, not all organic farms have quick access to manures and recycling is restricted by the prohibition of the utilization of sewage sludge due to current concerns over the introduction of potentially toxic elements, organic pollutants and disease transmission. Additionally, the present global market, food transported to large distances from the farm, leads to a big export of nutrients. Exported nutrients must be replaced to avoid nutrient depletion of soils. Nutrient budgeting suggests some cause for concern over the sustainability of organic systems due to their dependence on feedstuffs and bedding for inputs of phosphorus (P) and potassium (K), and on the very variable fixation by legumes or imports of manure or compost for nitrogen (N); pollution and net mineralization from soil reserves appear to comprise a large a part of the N supply on some organic farms. Losses of N from organic systems can also be as large as those from conventional systems and, being hooked in to cultivation and therefore the weather, they're even harder to regulate than those from fertilizers applied to standard farms. There's some evidence of P deficiency in soils under organic production, which is difficult to be replaced by K in farm produce. Organic farming systems could also be sustainable and have the potential to deliver significant environmental benefits, but these depend upon specific cropping and management practices on each farm. It is important that we study and improve nutrient management on all farm systems and within the context of plant, animal and human health so as to develop more sustainable farming systems.

5.2 NUTRIENTS APPLIED IN ORGANIC FARMING

Organic farming systems are very diverse and are found across the world, with a range of crop and animal enterprises often linked together. However, strong unifying principles link this wide range of systems and management practices. In many countries organic farming has a clear legislative basis and certification schemes for production and processing. Few nutrients are applied in organic farming sectors which are used as good fertilizer.

5.2.1 Nitrogen

Low intensity livestock systems, completely based on rough grazing and in which no attempt is made to increase N supply, are common in the uplands; they represented 43% of the land area converted to organic production in the UK in 1997 (MAFF, 1999). The remaining organic farms are mostly lowland livestock or mixed arable-livestock enterprises, with a substantial part (at least 50%) of the farm in a ley in any one year. In this type of system, N is supplied by biological fixation, usually through the inclusion of red and /or white clover in grass mixtures. Estimates of fixation in organically managed grass-white clover leys range from 100 to 235 kg N ha-1 (Vinther & Jensen, 2000). It is possible to use stockless systems. In other cases, organic farmers are permitted to use green cover containing more than 5% legumes in the seed mixture (MAFF 1999). Most common is a two-year grass-red clover ley, which can accumulate 250-370 kg N ha-1, and is cut and mulched to return N to the soil for subsequent crops (Stopes et al., 1996). Good growth of legumes also depends on adequate amounts of P and K in soils. Cultivation and subsequent mineralization are key to the release of this fixed N, but mineralization is difficult to control. Poor matching of soil supply and crop demand for N can lead to losses by leaching or denitrification or both (Philipps and Stopes, 1995; Lord et al., 1997). The transition from ley to arable cropping in an organic rotation is generally associated with the highest loss, with up to 180 kg N ha-1 leached in the winter after ploughing. Season, timing and intensity of cultivation have been shown to have a substantial effect on this loss. Cultivation of leys in spring, followed by spring cropping, reduces nitrate leaching markedly (Watson et al., 1993). However, where autumn sowing is an option, winter cereals tend to dominate, due to their greater profitability. The careful management of animal manure to minimise losses and optimise nutritional benefits is a key feature of stocked organic systems. At present there is no limit to the application of N in manures and slurries in organic systems, other than guidance provided by the Codes of Good Agricultural Practice.

5.2.2 Phosphorus and Potassium

In case of mixed and animal based systems, animal feeds and bedding represent relatively large inputs of P and K to organic farms (Fowler et al., 1993, Nolte and Werner, 1994). Any necessary additional P and K is applied strategically within the rotation, with one application of rock phosphate expected to supply P for a number of following crops. However, organic farming seeks to optimize the recycling of P and K and to keep imports as small as possible. In animal-based and mixed systems, good manure management is therefore essential. Manure and slurry are used to redistribute nutrients around the farm. Phosphorus dominantly occurs in organic forms in manures, while K is found in soluble forms; large leaching losses of K can therefore occur during manure storage and composting (Fowler et al., 1993).

5.2.3 Others Nutrients

It is found that liming is good practice in organic farming, using ground chalk, limestone and seaweed; slaked or quick limes are not permitted. Deficiencies of other minor nutrients are not widely reported in organic crops because trace amounts of these are often present in purchased feeds and approved amendments. Some concern has been expressed over sulphur and zinc requirements to achieve bread-making quality in wheat (Starling & Richards, 1993). There is also concern over trace element deficiencies in organic livestock (Weller et al., 1996; Andrews, 1999). However, little research has been carried out in this area. We have confined our study to N, P and K.

5.3 SUSTAINABILITY OF NUTRIENT SUPPLIES IN ORGANIC FARM SYSTEMS

5.3.1 Soil Analyses and Indices

Soils in England and Wales are routinely analysed for available P using Olsen's method (extraction with 0.5M sodium bicarbonate; Olsen et al., 1954) and for available K by extraction with ammonium nitrate (MAFF, 1973) or ammonium acetate (Metson, 1965). Depending on the content of available P and K, soils are grouped into indices from 0 to 9. Generally, soils at Index 0 and 1 are deficient and in conventional production the aim is to maintain an Index of 3 (MAFF, 1994). Indices of available P and K in England and Wales are measured in the Representative Soil Sampling Scheme (Skinner and Todd, 1998). The double lactate (calcium lactate + lactic acid) method is used for the analysis of P and K in soils from organic farms. For arable and grassland soils, lactate extractable P range between 100–300 mg kg^{-1} and for lactate extractable K remains between 100–200 mg kg^{-1} soil regarded as sufficient. These amounts cannot be easily related to conventional 'Available' P and K contents. However, we have made a limited comparison of organic and conventional soils using conventional P and K indices. Data for soils submitted for analysis to the Organic Advisory Service during 1997 were compared with those from conventionally-farmed soils. Some 39% of soils from organic farms were at an 'Olsen' P index of 0 or 1, 23% at Index 2 and 38% at Index 3 or above. Double lactate extraction of P and K showed that 86% of these soils were deficient in available P and 36% were deficient in available K. By contrast, for approximately the same years, only 15% of conventional arable soils were at an 'Olsen' P Index of 0 or 1, and only 20-25% of conventional grassland soils were at a P Index of 0 (Chambers et al., 1996). Some 30% of conventional arable soils were at a K index of 0 or 1, 50% at Index 2 and 20% at Index 3 or above.

5.3.2 Nutrient Budgets

Clark et al. (1998) showed that most changes in nutrient indices, such as those reported above, could be predicted from a consideration of the inputs and outputs of nutrients, i.e., the nutrient budget. Nutrient budgets have been compiled around the world, using a variety of scales and methodological approaches (Scoones and Toulmin, 1998; Watson and Atkinson, 1999). They measure or estimate the inputs and outputs of nutrients (usually N, P and K) to a field, farm or system, usually at the 'farm gate'. Farm gate budgets do not usually include the necessarily very detailed measurements of losses such as leaching, denitrification and ammonia volatilisation, consider each field separately, or measure transfers between fields. Nor do they provide information on soil processes or biological inputs and outputs of nutrients, which are particularly important for N.

5.3.3 Impact of Nutrient Losses

Measuring full nutrient budgets, including losses and internal flows, is essential for proper interpretation. An analysis of spatial flows can highlight the enrichment or depletion of nutrients from different parts of the farm. For example, the practice of spreading manure and slurry preferentially on fields nearest the manure store can lead to preferential enrichment at the expense of other fields (Bacon et al., 1990). The quantification of internal flows is also particularly important for organic farming because of the desire to maintain and improve the internal recycling of nutrients. However, as the complexity of the approach increases there is a need to measure or estimate increasing number of variables; either costs or errors increase. Below we estimate N, P and K losses from organic farms to enhance the budgets. Nitrogen. Lord et al. (1997) found no difference in N leaching from a number of comparable organic and conventional farms; all had an average loss of 50 kg N ha^{-1} yr^{-1} over a whole rotation. Using modelling approaches, the Organic Farming Study (OFS; Cobb et al., 1999) found losses from organic farms (52 kg N ha^{-1} yr^{-1}) to be two-thirds of those from conventional farms (78 kg N ha^{-1} yr^{-1}). The amount of nitrate leaching was related to the timing of cultivation, crop patterns and management. Few measurements of losses by ammonia volatilisation have been made. Losses of ammonia from the OFS mixed farm were estimated at c 25 kg N ha^{-1} yr^{-1}; estimated losses of 50-100 kg N ha^{-1} yr^{-1} on a nearby conventional mixed system. In general, losses are determined by stocking rates, which are lower for organic farms than for comparable conventional systems. We have used values of 25 kg N ha^{-1} yr^{-1} for the mixed upland farm and 50 kg N ha^{-1} yr^{-1} for the lowland dairy farm. There may be significant losses of ammonia from stockless systems where leys are cut and mulched, but no data are available so we have put such losses at zero.

5.4 CONCLUSION

The data presented here suggest some cause for concern over the sustainability of organic systems because of their dependence on feedstuffs and bedding for inputs of P and K, and on the very variable fixation by legumes or imports of manure or compost for N. Air pollution and net mineralization from soil reserves appear to comprise a large part of the N supply on some organic farms. Losses of N from organic systems can also be as large as those from conventional systems. Being dependent on cultivation and the weather, they are even more difficult to limit than those from fertilisers applied to conventional farms. There is some evidence of P deficiency in soils under organic production. However, with careful management of manure and the effective use of legumes and by using permitted inputs for P and K, organic farms can be managed sustainably.

Long-term Trends

Care must be taken when reaching conclusions about the sustainability of farming from farm gate nutrient budgets. Applications of P and K are often made during the ley phase of an organic rotation to supply the whole rotation. Thus, on the lowland dairy farm, rock phosphate is used occasionally to redress the small P deficit. This does not appear in its nutrient budget because the P was not applied in the year the budget was compiled. An annual budget may also mask important differences between fields. Several years' data or preferably a whole rotation, should really be considered. The weather also has an important effect through its impact on losses. Nitrogen budgets have been calculated for the Broadbalk Continuous Wheat Experiment at IACR-Rothamsted, now in its 157th year. On this experiment, all the inputs and outputs of N to the various plots, including the accumulation or depletion of soil reserves, can be measured or calculated. From 1990-98 no single year produced a balanced budget. Only for the full 8-year period, when the mean rainfall equalled the long-term average, could a balanced budget be obtained. In dry years a net surplus of inputs over outputs was observed, with the reverse in wet years. This can be explained by net immobilization of N in soil organic matter in dry years and net mineralization in wet years leading to large leaching losses. The deficits seen in the OFS and in the organic upland farm above, and the surplus in the stockless arable farm, may therefore be temporary, balanced by corresponding surpluses and deficits in other years as on Broadbalk. Nutrient cycling at the wider scale organic produce is sold to the national and international markets that dominate food production, leading to a net export of plant nutrients from most organic farms. We have shown that organic farms remain sustainable only through the use of bought-in feed and bedding, composts and, in some cases, air pollution. UKROFS rules state that, ideally, all imported feed should be from feedstuffs produced to UKROFS standards. However, at present between 10 and 30% of the dry matter, on a daily basis, may be imported from nonorganic sources, depending on the type of animal. At a wider scale there are issues concerning the import of nutrients in fertilising materials over long distances. The concept of food miles, that is, the distance travelled by food products between production and consumption, is now in common use. There is a similar issue in relation to sustainability in terms of 'resource miles' or the distance travelled and energy costs associated with freight of products such as rock phosphate being brought from as far away as Tunisia and Morocco to the UK for use on organic farms. In the long term, organic farming standards may need to consider more fully the use of locally sourced by-products and waste materials for nutrient supply. The ideal of sustainable organic farms that do not consume nonrenewable resources or pollute the environment will remain elusive while the global market economy dominates agriculture.

5.5 QUESTIONS

Long Questions

1. Briefly describe restrictions to nutrient use in organic farming.
2. Rules, regulations and limitation of micronutrient application in organic farming in Indian and global perspective.
3. How effectiveness acts on nutrient cost applied organic farming?

Short Questions

1. How many different kinds of micronutrient can be used in organic farming.
2. How nutrient budgets should be maintained in organic farming.
3. Write a short note on micronutrients used in organic farming.

REFERENCES

Andrews A. 1999. Vitamin e and selenium. Organic farming, summer 1999, p30. MAFF (Ministry of Agriculture, Fisheries and Food). 1973. *The Analysis of Agricultural materials. Technical Bulletin,* 27. HMSO, London.

Bacon, S. C., Lanyon, L. E. & Schlauder R. M. 1990. Plant nutrient flow in the managed pathways of an intensive dairy farm. *Agronomy Journal,* 82: 755-761.

Chambers B. J., Webb J., Mitchell, R. & Garwood T. 1996. Soil nutrient status and ph-fertility indicators. Indicators of soil quality. The seventh silsoe link conference, Silsoe, 28 February 1996.

Clark M. S., Horwath W. R., Shennan C. and Scow K. M. 1998. Changes in soil chemical properties resulting form organic and low-input farming practices. *Agronomy Journal,* 90: 662-671.

Cobb D., Feber R., Hopkins A, Stockdale L., O'riordan T., Clements B., Firbank L., Goulding, K., Jarvis, S., and Macdonald, D. 1999. Integrating the environmental and economic consequences of converting to organic agriculture: Evidence form a case study. *Land Use Policy,* 16: 207-221.

Fowler S. M., Watson C. A., Wilman, D. 1993. N, P and K on organic farms: Herbage and cereal production, purchases and sales. *Journal of Agricultural Science,* Cambridge 120: 353-360.

Lord E., Stopes C and Philipps L. 1997. Assessment of relative nitrate losses from organic and conventional farming systems, based on recent measurements. Final report to Maff, contract of 141. UK ADAS, Wolverhampton.

MAFF (Ministry of Agriculture, Fisheries and Food). 1973. The Analysis of Agricultural Materials. Technical bulletin 27. HMSO, London. MAFF (Ministry of Agriculture, Fisheries and Food). 1994. *Fertiliser Recommendations for Agricultural and Horticultural Crops,* 6th edn. Reference book 209. HMSO, London.

MAFF (Ministry of Agriculture, Fisheries and Food). 1999. Review of the Organic Aid Scheme. MAFF, London.

Metson, A. J. 1965. Methods of chemical analysis for soil survey samples. Soil Bureau Bulletin 12, New Zealand Department of Scientific and Industrial Research, Wellington, p. 117.

Nolte, C. and Werner, W. 1994. Investigations on the nutrient cycle and its components of a biodynamically-managed farm. *Biological Agriculture and Horticulture,* 10: 235-254.

Olsen S. R., Cole C. V., Watanabe F. S. and Dean L. A. 1954. Estimation of Available phosphorus in soils by extraction with sodium bicarbonate. USDA circular 939.

Philipps L. and Stopes C. E. 1995. Organic rotations and nitrate leaching in the UK. *Biological Agriculture and Horticulture,* 11: 123-134.

Skinner, R.J. and Todd, A.D. 1998. Twenty-five years of monitoring ph and nutrient status of soils in England and Wales. *Soil Use and Management,* 14: 162-169.

Stopes C., Millington S. and Woodward L. 1996. Dry matter accumulation by three leguminous green manure species and the yield of a following wheat crop in an organic production system. *Agriculture, Ecosystems and Environment,* 57: 189-96.

Scoones, I. And Toulmin, C. 1998 Soil nutrient balances: what use for policy? *Agriculture Ecosystems and Environment,* 71: 255-267.

Starling W. and Richards M. C. 1993. Quality of commercial samples of organic ally grown wheat. *Aspects of Applied Biology,* 36: 205-209.

Vinther, F.P. and Jensen, E.S. 2000. Estimating legume N2 fixation in grass-clover Mixtures of a grazed organic cropping system using two 15n methods. *Agriculture, Ecosystems & Environment,* 78: 139-147.

Watson C.A., Fowler, S.M. and Wilman, D. 1993. Soil inorganic-n and nitrate Leaching on organic farms. *Journal of Agricultural Science,* Cambridge 120: 361-169.

Watson C. A. and Atkinson D. 1999. Using nitrogen budgets to indicate nitrogen use efficiency and losses from whole farm systems: a comparison of three methodological approaches. *Nutrient Cycling in Agroecosystems,* 53: 259-267.

Weller R.F., Cooper A. and Wilkinson R. 1996. Animal health. In: conversion to organic milk production. Iger technical review no. 4. (eds. Haggar, R. and Padel, S.), pp. 96-111.

CHAPTER 6 Organic Production Requirements

6.1 INTRODUCTION

Organic crop production system is one of the new frontiers of agriculture in a eco-friendly way. But one of the biggest challenges is to minimize the loss of produce from pest and to maintain the soil health without using any synthetic chemical. So, one producer should gain proper idea and knowledge about the production system and its procedure. The new producer should tour to an organic farm prior to its own production (Meena et al., 2013). This chapter highlights the requirements of organic production system and its standards.

6.2 CHOICE OF CROPS AND VARIETIES

Seeds or any other planting materials should be collected from a certified organic grower. The selected variety or species should adapt to the environment, soil and be resistant to pests. It is advised to use organic seed and planting material. But if, certified organic and planting material is not available, grower can use chemically untreated conventional planting materials. Any planting materials which are genetically modified or transgenic plants should not be allowed for use.

6.3 DURATION OF CONVERSION PERIOD

A conversion period is required for the organic system to build soil fertility and establishment of organic system. In this period, all required actions are started to reach the goals. The duration of conversion period depends on the ecological situation and the land history.

A conversion period of at least 2 years before sowing or 3 years before the harvest of the product is required to meet the standards of organic crop production system in case of perennial crops. It is upon the accredited inspection and certification agency to extend or reduce the conversion period in certain cases like no land use for 2 years or more than that. The certification agency allows the grower to sell his products by labelling "produce of organic agriculture" or similar to it. The feed produced in the farm in the first year of conversion can be defined as organic, only to feed the animals reared inside the farm and not to be sold in market as 'organic'.

6.4 DIVERSITY IN CROP PRODUCTION

Crop production can be diversified to gardening, farming and forestry by considering the structure and fertility of the soil as well as the quality of the ecosystem. In this diversification the nutrient and mineral loss should be less.

Diversification can be possible with crop rotation by using legumes or any other crops. During diversification the soil should be covered properly with diverse type of plant species to minimize the loss. Diversification in crop rotation should be planned in a way that pest and weed population will be less, soil organic matter will be high, increase of the activity of beneficial microorganisms and to increase fertility and soil health.

6.5 FERTILIZATION

In organic production system, maintaining soil health and restoring soil nutrients is one of the major challenges. To restore the fertility, biodegradable materials such as materials from plant and animal origin as well as microorganism derived should be used which may increase the fertility or will help to maintain it by increasing the biological activity. This is the best way of fertilization in organic production system.

Fertilization system should focus on:

- Leaching loss of nutrients and to minimize it.
- Prevention of accumulation of heavy metals and other pollutants.
- Maintaining of proper pH in soil.
- Use of non-synthetic mineral fertilizers as the supplementary.

The standards set by the certification agency for the fertilization programme are as follows:

- The basis of fertilization should be microbial origin or animal or plant derived materials.
- The agency can set the standards on the use of animal waste as its overuse is one of the ways to pollute the environment.
- Standards can be set up by the agency on the huge of amount of biodegradable material plant animal or microbial origin depending upon the nature of the crop as well as the local conditions.
- Human excreta should not be used in fertilization programme.

- Mineral fertilizers are allowed with their natural composition and not permitted after any chemical treatment (Table 6.1).
- Use of any synthetic nitrogen as fertilizer is prohibited.
- Restrictions should be imposed for the use of magnesium fertilizers, trace elements, manures, mineral potassium and other substance containing heavy metals like basic slag, sewage, sludge, rock phosphate, etc.

Table 6.1 Restricted and permitted products for soil management

S. No.	Location of matters produced	Types of matter	Permitted	Restricted	Not allowed
1.	Inside organic farm unit	Crop residues or green manure, straw or other mulch	✓		
2.		FYM, slurry, urine, poultry manure	✓		
3.	Outside organic farm unit	FYM, slurry, urine, poultry manure		✓	
4.		Fish and fish products without preservatives		✓	
5.		Blood meal, meat meal without preservatives		✓	
6.		Human excreta			✓
7.		Saw dust and wood shavings	✓		
8.		Seaweeds after physical process		✓	
9.		Peat without synthetic additive	✓		
10.		Straw		✓	
11.		Vermicast		✓	
12.		Compost from plant residue	✓		
13.		Compost from organic household		✓	
14.		Basic slag		✓	
15.		Pulverised rock		✓	
16.		Calcium chloride		✓	
17.		Microbiological preparations		✓	

6.6 PLANT PROTECTION

Organic farming system should focus to minimize the loss of the produce by reducing the insect, disease and weed infestation. It can be achieved by the use of fertilization, green manuring, crop rotation, companion planting, selection of variety and crop and other natural methods. Use of synthetic chemicals is restricted in this system.

Pests and weeds can be controlled by following different cultural, physical, mechanical and biological methods. Examples of such methods are heat treatment, soil solarization, application of natural predators, crop rotation, etc.

Some standards for the use of products to control the pest by certification agency are given below:

- The allowed products for control of paste and weeds are from local plants, animals or microorganism derived.
- The products for use should be checked and accepted by the agency. Branded products require an evaluation for use in the field.
- Any type of heat treatment and physical methods are allowed for pest and weed management.
- Any equipment and implement which is normally used in conventional fields need to be cleaned properly prior to use in an organic field.
- Any type of synthetic chemical is not permitted for using in the organic field. But some permitted chemicals can be used to certain extent (Table 6.2).
- Any type of genetically modified organisms or products cannot be used.
- Use of any synthetic growth regulator and synthetic dyes are also prohibited.
- All the above measures and methods should meet the standard shared by the certification agency.

Table 6.2 Restricted and permitted products for soil management

S. No.	Types of substances to use for plant protection	Name of the substance	Permitted	Restricted	Not allowed
1.	Substances from plant and animal origin	Neem oil, neem extract		✓	
2.		Gelatine	✓		
3.		Plant based extracts	✓		

4.		Propolis		✓	
5.		Release of parasite or predators of insect pest		✓	
6.		Tobacco tea			✓
7.		Casein	✓		
8.		Seaweeds and its extract		✓	
9.		Mushroom extract	✓		
10.		Clay	✓		
11.		Lime sulphur		✓	
12.		Sodium bicarbonate	✓		
13.		Mineral powders			✓
14.	Minerals	Diatomaceous earth		✓	
15.		Traps	✓		
16.		Ethyl alcohol			✓
17.		Microbiological preparations	✓		

6.7 CONTROL FROM CONTAMINATION

The field should be cleaned and sanitized properly to check any type of contamination within or from outside. The certification agency set some limits for the application of heavy metals. Maximum accumulation of pollutants and other heavy metals are not allowed in organic field.

The certification agency can analyse the product or the soil and water to determine the accumulation of heavy metals or any other pollutants. In case of protected cultivation, products from polyethylene and polypropylene or polycarbonates are allowed. Polychloride based products are prohibited from use (Hole et. al., 2005).

6.8 SOIL AND WATER CONSERVATION

The soil and water should stand the standards set by the certification agency. Soil and water should be free from any type of pollution, erosion loss or salination. Here water refers to both surface and ground level.

The standards set by the certification are as follows:

- Clearing of forest prior to farming in the same place is prohibited.

- Measures should be taken to minimise the erosion loss of water and soil.
- Clearing of land through burning of straw or crop residues is restricted to some extent.
- Measures should be taken to prevent the soil and water from salination.
- Soil and water should contain the minimum amount of heavy metals or pollutants.

6.9 COLLECTION OF NON-CULTIVATED MATERIALS FROM PLANT ORIGIN

From the stage of collection to marketing, measures should be taken to maintain the quality and sustainability of the product as well as the ecosystem. Products from forest can only be certified if collected from a sustainable environment. The collection area has the minimum appropriate distance from any type of pollution and contamination area or conventional farming (Finamore et al., 2004).

6.10 CONCLUSION

The certification and inspection agency set the standards parallelly with the minimum requirements by the central government. The grower should follow all these standards and the procedures. After fulfilling all the criteria, the producer and its product will get an organic certificate for marketing. The grower should gain basic knowledge on the organic crop requirements, the restricted and allowed substances, the soil and water quality, etc. By knowing and following all these regulations, he can be an organic certified grower (Nanda et al., 2003).

6.11 QUESTIONS

Long Questions

1. What are the requirements in organic farming system? Explain in brief?
2. List the restricted and permitted substances in organic crop system?
3. How to manage the soil health and quality in organic system?
4. How to control the pest and weed infestation in organic system?
5. What do you mean by conversion period and how it helps in organic system?

Short Questions

1. Restricted substances in organic system
2. Conversion period
3. Organic agriculture
4. Pest management in organic system
5. Field sanitization

REFERENCES

Finamore A, Britti MS, Roselli M, Bellovino D, Gaetani S, Mengheri E. (2004). Novel approach for food safety evaluation. Results of a pilot experiment to evaluate organic and conventional foods. *Journal of Agricultural and Food Chemistry*, 52: 7425-7431.

Hole DG, Perkins AJ, Wilson JD, Alexander IH, Grice PV, Evans AD (2005). Does organic farming benefits biodiversity. *Biological Conservation*, 122: 113-130.

Meena, R.P., Meena, H.P. and Meena R.S. (2013). Organic farming: Concept and components, *Pop. Kheti*, 1(4): 5-14.

Nanda SK, Bhattacharya B, Chandra PS (2003). Promoting organic farming in India: Extension implications. *Management Extension Research Review*, 4(2): 9-20.

Unit 3

Bio-intensive Nutrient Management and Biofertilizers

7 CHAPTER Choice of Crops and Varieties in Organic Farming

7.1 INTRODUCTION

The demand for organic agriculture is expanding day by day. Consumers are also shifting from non-organic products to organic products due to their eco-friendly nature and no chemical residues. The high demand of organic produce has simultaneously increased the producers to continue the supply chain. But in organic system one of the biggest challenges is to increase the yield with low or no use of chemicals for soil health management as well as pest control. Therefore, one of the important steps in organic agriculture is selection of the right crop and variety which can minimize the loss. This chapter describes the selection process of right crop and variety and factors affecting it.

The ultimate goal of the producer should be improving the productivity with the available infrastructure and it can be achieved through:

- Proper use of natural resources available in the locality
- Adopting the low-cost technologies
- Adopting appropriate farming system
- Producing quality foods
- Following time management practices
- Getting practical knowledge with scientific technologies

At the time of selection of crop and variety the following points should be noted by the producer.

- All the seeds and planting materials should be collected from a certified producer.
- If organic seed or planting material is not available, then the producer can use chemically untreated conventional seed or planting material available locally.
- Genetically engineered seeds or transgenic plant or plant material are prohibited.

- The producer should check if the selected crop and variety is adapted to the specific soil and climatic condition or not, resistant against pest and weeds, genetic modified or not, etc.

7.2 CROP SELECTION

In organic farming system crop selection is one of the major steps. When a producer is selecting a crop, he has to focus on adaptability of the crop in the land, climatic factor, resistant to insect pest, no genetically modified or transgenic plant, marketability, etc. These are explained below (Fig. 7.1).

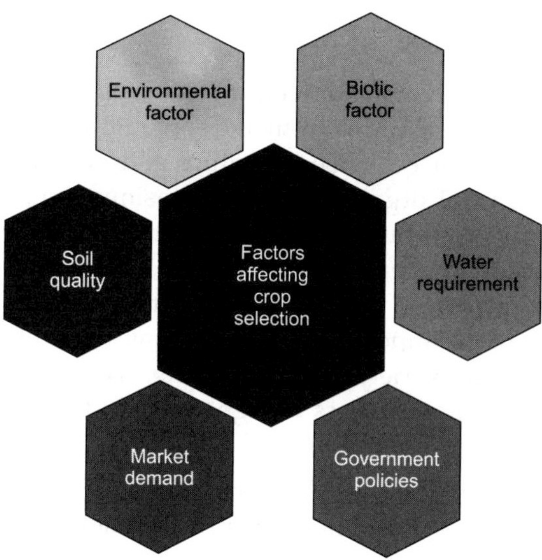

Fig. 7.1 Factors affecting crop selection in organic system.

7.3 ENVIRONMENTAL FACTORS

Environmental factors refer to rainfall, temperature, relative humidity, average hour of sunlight and frequency of extreme weather conditions. Therefore, when selecting a crop, it should be adapted do the above-mentioned local environmental factors.

7.4 SOIL CONDITION

Prior to growing of a crop, a producer should test the soil to know the organic matter content, pH or any other heavy metals in the soil. The crops depend on the soil health and quality such as organic matter content, soil texture, soil colour, porosity, pH, fertility, etc. But this can be altered or modified by adding manure and fertilizers which are permitted to use with restricted limits.

7.5 BIOTIC FACTORS

Biotic factors normally refer to the living organisms such as pathogens, insects, weeds or some beneficial microorganisms living in the soil. So, before choosing the crop, the grower should properly check the crop to be selected which can be adapted or resist to the biotic factors of the locality. For this, he can take the help of extension workers, nearby farm or any other agriculture officer of the same locality. If the crop selection is wrong, then the yield will be reduced highly and cost of cultivation will increase.

7.6 MARKETABILITY

The selected group should have high marketability in the specific location. The crop should be trendy and can cover a large group of interested consumers. For example, in vegetables like tomato, brinjal cauliflower, cash crops, or mushrooms can be preferred. Crops should be chosen in a way that they return high profit with less inputs.

7.7 AVAILABLE TECHNIQUES IN THE LOCALITY

Crops should be chosen in a way that they don't need any extra implements and can be managed with the existing ones. If the implements are available in the locality or neighbouring farm then the producer can go for the crop. Aside the planting varieties which require intensive care and specialized implements. There should be year-round production or continuity of chain of supply to the market. Otherwise the producer can choose the crop which can be stored for a long term without any spoilage. For example, grains, cereals, root crops, dry beans, etc.

7.8 WATER REQUIREMENTS

The crops should be selected on the basis of water requirements. If the location receives high rainfall, then the farmer can go for the crops which require high water. If less rainfall, but still farmer has proper facilities of tanks, wells, dams or any other micro irrigation system or pumps set with no or less electric disturbance, then also he can go for high water required crops.

7.9 CROPPING PATTERN

If the farmer wants intercropping or mixed cropping, relay cropping, multistoried cropping, crop rotation, then he should have much area, proper knowledge of cropping pattern. If he doesn't have proper idea, then he can take suggestions from experienced growers or extension workers.

7.10 ECONOMICS AND RISK FACTOR

The selected crop or cropping pattern should be suitable for the growing area. This should not require any extra implements or extra resources. Before executing the cropping pattern, the grower should analyse the risk factor such as how much profit, how much loss, how to manage the problems, is it ok to take risk in this much profit, etc.

7.11 LABOUR AVAILABILITY AND MECHANIZATION

The grower should focus on the following questions prior to executing:

- Is this cropping system could be managed through their family labour?
- If not, then are farmers available in the locality or not?
- Are proper machines and implements available in the locality or not?
- Are the available machinery cost effective and affordable or not?
- Are the available labour has the potential and can handle the available machineries?

7.12 POST-HARVEST STORAGE AND PROCESSING

Crops and cropping patterns should be selected in a way that there will be proper storage facility in the locality. Harvested products can easily be transported to the nearby market. There should be casualty of primary processing, technologies available for value addition. The growers should have proper knowledge about the standards for harvesting and processing of the product.

7.13 GOVERNMENT POLICIES AND SCHEMES

The selected crops should favour the government policies – an incentive can be claimed from the government. The grower or farmer should have proper knowledge about the new agricultural technologies. For this he should have proper access to kitchen call centre, nearby KVK, agricultural technology management agency or any other extension workers. Proper Contacts should be maintained with the NGOs, agri-clinics or agri-business centres, government agricultural officers to get update with the recent schemes and technologies.

7.14 CONCLUSION

Many farmers and growers are now considering organic farming as an interesting and trendier one. Likewise, the consumers for organic products are increasing day by day. Therefore, to maintain the continuity of supply chain in the market, the growers should make a proper strategy and do accordingly. He must have back-up plans for all type of risk factors. Selection of crop and variety is the most important step which is dependent on a wide range of factors. Therefore, proper measures should be taken by the grower.

7.15 QUESTIONS

Long Questions

1. What are the areas a producer should focus at the time of selection of a crop?
2. What are the factors affecting in selection of a crop?
3. Explain the role of government and non-governmental organizations in selection of a crop or cropping pattern. How can they can help a grower regarding this?
4. What are the risk factors a farmer or grower can face in organic farming system?
5. Explain the biotic factors and environmental factors affecting the organic production system?

Short Questions

1. Biotic risk factors in organic farming
2. Climatic factor in organic system
3. Crop rotation
4. KVK
5. Agri-clinics
6. Post-harvest processing of organic products

REFERENCES

APEDA (Agricultural and processed food products export development authority). (http://www.apeda.gov.in/apedawebsite/organic/Organic_Products.htm#:~:text=The%20APEDA%2C%20Ministry%20of%20Commerce,organic%20farming%20and%20marketing%20etc.)

Bruschi V, Shershneva K, Dolgopolova I, et al. (2015). Consumer perception of organic food in emerging markets: evidence from Saint Petersburg, Russia. *Agribusiness*, 31(3): 414–432.

Illukpitiya P, Khanal P. (2016). Consumer perception of organic food and product marketing. *Org Farm Sustain Agric*, 9: 315–324.

Moonen A-C, Barberi P. (2008). Functional biodiversity: an agroecosystem approach. *Agric Ecosyst Environ*, 127: 7- 21.

National Programme for organic production, Department of Commerce, Ministry of Commerce and Industry, New Delhi. 6th edition, 2005.

8 CHAPTER

Biological Intensive Nutrient Management: Organic Manures, Vermicomposting, Green Manuring, Recycling of Organic Residues

8.1 INTRODUCTION

The dramatic expansion of agriculture and application of chemical fertilizers is increasing day by day worldwide. In previous years, many negative impacts have been observed in our environment such as loss of biodiversity, disturbances in food chain, climate change, etc. The heavy use of conventional agro-chemicals is a red alert for our future generations because of their residual effect in soil for long periods. Sustainable agriculture is the only one solution to reduce the threat (Dahal et al., 2016). Sustainability in agriculture can only be achieved by biological intensive nutrient management system which includes all possible risk-free methods of agriculture such as organic manuring, vermicomposting, green manuring, etc.

8.2 BIOLOGICAL INTENSIVE NUTRIENT MANAGEMENT

It is a biological and ecological approach for management or manipulation of soil nutrients to reduce the risk of chemicals in all possible ways. It can be achieved by the use of organic manures, vermicomposting, green manuring and recycling of organic wastes. It is an old concept which balances the agri farming system (Ndegwa et al., 2000).

8.3 ORGANIC MANURE

The use of organic manure is one of the key factors for eco-friendly nutrient management of soil. It enriches the soil with various minerals, nutrients by improving the chemical, physical and biological properties of soil. It also accelerates the growth and spread of soil microorganisms and decomposers. During decomposition, organic manures are released and mixed with minerals which is taken by the plant. To improve the poor soils, organic manuring is the best suitable method. Manures are of plant, animal based origin containing nutrients in complex organic form.

On the basis of concentration, organic manure is of 2 types: concentrated organic manure such as animal originated or plant originated manure, and bulky organic manure such as farm yard manure, green manure, sewage waste, vermicomposting, sheep manure, etc (Figure 8.1).

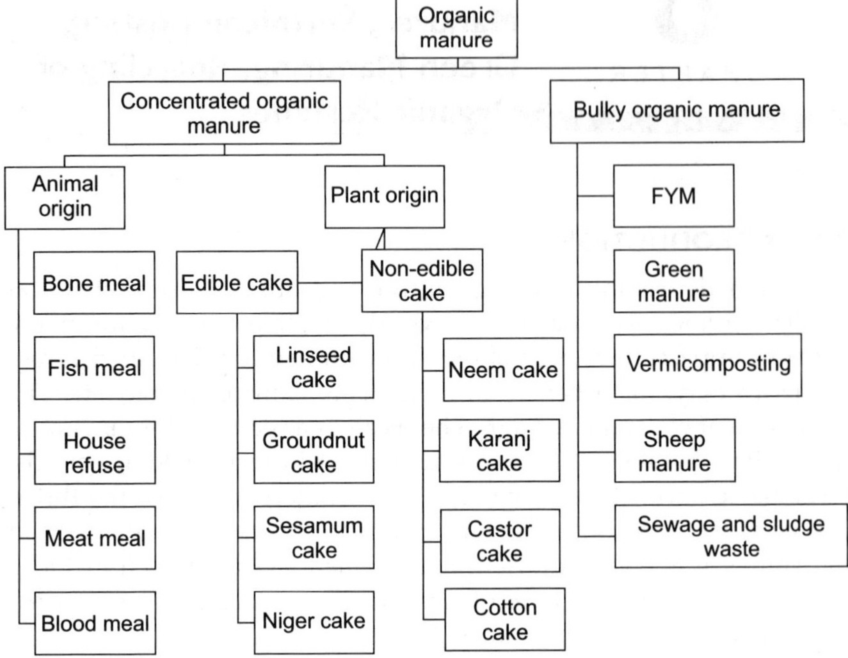

Fig. 8.1 Classification of organic manure.

8.4 ADVANTAGES OF ORGANIC MANURES

- High organic matter
- Improves physical, chemical and biological property of soil
- Improves growth of plant by releasing growth promoters
- Eco-friendly and cost effective
- Recycle the materials
- Balance the carbon and nitrogen ratio in the soil
- Increase soil fertility
- Improve the water holding capacity of soil
- Improve soil structure and texture
- Enrich soil with many other essential elements and minerals
- Suppress growth of weed and pathogens
- Activate beneficial microorganisms for nutrient enrichment
- Slow release of nutrients prevents reduction loss and make them available for long time

Table 8.1 Organic manures and their nutrient content

Sl. No.	Manure	Nitrogen (N)	Phosphoric acid (P_2O_5)	Potash (K_2O)
1.	Raw bone meal	3-4	20-25	-
2.	Fish meal	4-10	3.9	0.3-1.5
3.	Blood meal	10-12	1.2	1.0
4.	Press mud	1-1.5	4-5	2-7
5.	Coir pith	1.2	1.2	1.2
6.	Rural compost	0.5-1.0	0.4-0.8	0.8-1.2
7.	Urban compost	0.7-2.0	0.9-3.0	1.0-2.0
8.	Dry sewage sludge	2.0-3.5	1.0-5.0	0.2-0.5
9.	Farm yard manure	0.4-1.5	0.3-0.9	0.3-1.9
10.	Sugarcane waste	0.35	0.10	0.60
11.	Tobacco	1.12	0.84	0.80
12.	Groundnut husk	1.6-1.8	0.3-0.5	1.1-1.7
13.	Arhar	1.10	0.58	1.28
14.	Paddy	0.36	0.08	0.71

8.5 GREEN MANURES

Green manuring can be defined as a practice of incorporating or ploughing of green manure crops when the crops are in green stage or just after flowering stage. In green stage or before flowering, they carry more amounts of nutrients (Bond and Grundy 2001). It plays an important role in addition of organic matter, improving the soil fertility as well as soil's physical structure. Additionally, the leguminous green manure crops carry nodule bacteria in their root which helps in nitrogen fixation. They also take care of leaching and erosion loss. According to the green manure crop, they vary in nutrient content which is described in Table 8.2 (Dahal et al., 2016).

Table 8.2 Green manure crops and their nutrient content

S. No.	Green manure crop	Nutrient content (% dry weight basis)		
		Nitrogen	Phosphorus	Potassium
1.	*Sesbania aculeata*	3.3	0.7	1.3
2.	*Sesbania speciosa*	2.7	0.5	2.2
3.	*Crotalaria juncea*	2.6	0.6	2.0
4.	*Pongamia glabra*	3.2	0.3	1.3
5.	*Calotropis gigantea*	2.1	0.7	3.6
6.	*Azadiracta indica*	2.8	0.3	0.4
7.	*Tephrosia purpurea*	2.4	0.3	0.8
8.	*Gliricidia maculata*	2.9	0.5	2.8

8.5.1 Advantages of Green Manures

- They don't have any residual toxicity so referred to as one of the eco-friendly methods in agriculture.
- Improve the chemical and physical properties of soil.
- Suppress the growth of weeds and pathogens.
- Activate the growth of beneficial microorganisms.
- Nutrient leaching loss can be minimized by cultivating green manure crops.
- Cost effective.
- Use of nitrogenous fertilizer can be minimized as most green manure crops are leguminous.

8.6 COMPOST

The organic manures derived from organic waste through the activities of microorganisms under a controlled condition is known as compost. It is a method of solid waste management where the solid wastes are decomposed biologically in a controlled condition to store and apply it in the field in an eco-friendly way. Composting aims to stabilize and store the organic matter of agriculture, industry or kitchen waste to develop a slow release and potential fertilizer for improvement of soil nutrients, soil structure as well as soil health. In addition to this, it can also suppress the growth of weeds and pathogens.

Direct application of fresh crop residues in soil leads to immobilization of nitrogen highly. Therefore, composting is one of the alternatives to balance the carbon and nitrogen ratio. Composting can be done by using both urban and rural waste like straw, leaves, night soil, city garbage, kitchen waste, etc. (Rigby and Caceres 2001).

8.6.1 Composting Process

The process of composting is an aerobic method. It includes several techniques for conversion of organic waste to nutrient-rich fertilizer and soil conditioner by using several beneficial microorganisms (Table 8.4). The entire procedure is described in Table 8.3. The techniques are described below:

Windrowing: The waste material is cut in small size to form a windrow of 3 metre width, 1.2 metre height.

Inoculation: Required microorganisms are inoculated over the waste by sprinkling 4 kg of solid culture per tonne of waste and 2 kg liquid culture per tonne of waste.

Aero tilling: The waste should be aero tilled by using Aero tiller mixing for uniform mixing and proper oxygen availability for the microorganisms.

Application of spent wash: Nutrient-rich spent wash can be applied on windrows to maintain proper climatic condition. If it is not available, cow dung wash, a kitchen wash or vermi wash can be sprayed.

Table 8.3 Procedure of composting

Week	Activities
1st week	• Collection of waste • Cutting and windrow formation • Aero tilling • Inoculation of microorganisms
2nd-8th week	• Application of spent wash and aero tilling • Monitoring
9th – 11th week	• Enrichment of nutrients in compost

Table 8.4 Microorganisms involved in composting

Types of Micro-organism	Name	Dose of solid culture	Dose of liquid culture
N-fixing bacteria	*Azotobacter, Azospirillum*	2 kg/tonne	1 lt/tonne
P-solubilizing	*Bacillus polymixa*	4 kg/tonne	2 lt/tonne
K- mobilizing	*Fraturia aurantia*	4 kg/tonne	2 lt/tonne
Bio-control agents	*Pseudomonas fluorescence, Trichoderma viride*	2 kg/tonne	500 mL/tonne

8.6.2 Factors Influencing the Biocomposting Process

The various factors described below play an important role in composting process:

(a) **Moisture:** Moisture quantity accelerates the decomposition process. The required amount of moisture for composting is 50-60%. The slight change of moisture content leads to slow down the activity of microorganisms.

(b) **Aeration:** Proper aeration is an important part as the activity and growth of microorganism depends on the available oxygen quantity.

(c) **Nutrients:** High nutrient rich like agriculture waste, kitchen waste are mainly advisable. Waste materials carrying heavy metals and toxic wastes should be avoided.

(d) **Size of particles:** Less the size of particle, more the microbial growth. Hence, increasing the process of composting.

8.7 VERMICOMPOSTING

The method of composting by using the vermi (earthworm) is known as vermicomposting. Vermicomposting plays a crucial role in decomposition of organic matter, soil improvement and to enhance the quality of agricultural produce. In this method, various solid and liquid wastes from different industries like paper mills, sugar industry, food processing units can be used (Gupta and Garg 2008). Vermicasting is another important method which is mainly applied in the first layer of mulching with organic debris. Vermicasting is the compost carrying excreta of earthworm and their cocoon. These are rich in organic matter and plant nutrients. This accelerates the process of compost which can be increased by adding dung and other organic wastes as source of food for earthworm.

8.7.1 Process of Vermicomposting

For the process of vermicomposting, animal dung, cane trashes, poultry farm waste, city waste is collected and placed in a compost pit covered with green shade net to prevent sunlight. The pits should be of a height of 1-2 feet. Optimum moisture of 50-60% is maintained through water spraying or drip irrigation. Good quality and potential strains of earthworm is released in the compost pits. It takes 60-90 days to complete. This time period also depends on the number and strain quality of the earthworm and the environmental conditions. After completion of this process, earthworms are collected and reinoculated again in a new compost pit (Alvarez et al., 2001).

8.7.2 Advantages of Vermicomposting

- Greater and faster humification and nitrogen mineralization
- Contains hormone-like auxin which accelerates the growth of plant
- Decreases the amount of heavy metals in soil
- Eco-friendly
- Cost effective
- Proper balance between carbon and nitrogen ratio
- Stable and non-toxic soil conditioner and fertilizer
- Waste minimization in bulk density

8.8 FARM YARD MANURE (FYM)

The waste materials like cow dung or urine, fodder waste or roughages when decompose by the microorganisms, it is known as farm yard manure (FYM). It carries 0.5% nitrogen, 0.2% P_2O_5 and 0.5 % K_2O. Cow urine holds nitrogen in the form of urea which leads to volatilization loss. The composting pits should not contain any plastic or unnecessary materials to avoid contamination. The quality of FYM mainly depends on the factors cited below:

- Type of waste materials
- Type of animal
- Environmental factors
- Type of animal feed
- Storage, etc.

Methods to reduce the leaching loss

- Use of trench method
- Adoption of gobar gas plant
- Field management
- BYRE method for collection of FYM
- Covered method for storing FYM

8.9 CONCLUSION

Intensive nutrient management by altering or recycling the various waste materials of our surroundings is one of the best ways to sustainable agriculture. It uses different microorganisms, earthworms and crops which increase the nutrient and mineral uptake in an eco-friendly way (Curry 2002). But still more research is needed to increase the amount of nutrients and more microorganisms should be identified to use them in this novel procedure. Awareness among farmers is still lacking. Therefore, government should take steps to increase the knowledge of sustainable nutrient management by the help of NGOs and other organizations.

8.10 QUESTIONS

Long Questions

1. What do you mean by organic manure? Classify and list out the different types of manure with their nutrient composition?
2. What is vermicomposting? Explain the advantages of using earthworms in vermicomposting. Describe the process.
3. What is green manure? What will be the criteria to choose a crop for green manuring?

4. How different types of urban wastes can be recycled into an agriculture friendly product? Explain.
5. Briefly describe the process of composting and the factors affecting it. List out the microorganisms involved in it.

Short Questions

Write Short Notes on:

1. FYM
2. Green leaf manuring
3. Composting
4. Bulky organic manure
5. Concentrated organic manure
6. Vermicasting
7. Leaf composting
8. Biogas
9. Organic manure
10. Sustainable agriculture
11. Biofertilizer
12. Integrated nutrient management

REFERENCES

Alvarez, T., Frampton, G.K., Goulson, D., 2001. Epigeic Collembola in winter wheat under organic, integrated and conventional farm management regimes. *Agriculture Ecosystems & Environment,* 83: 95-110.

Bond, W., Grundy, A.C., 2001. Non-chemical weed management in organic farming systems. *Weed Research,* 41: 383-405.

Curry, D., 2002. Farming and Food: A Sustainable Future. Report for the Policy Commission on the Future of Farming and Food. DEFRA, London.

Dahal, K.R.; Sharma, K.P.; Bhandari, D.R.; Regmi, B.D.; Nandwani, D. 2016. Organic Agriculture: A Viable Option for Food Security and Livelihood Sustainability in Nepal. In: *Organic Farming for Sustainable Agriculture. Sustainable Development and Biodiversity;* Nandwani, D., Ed.; Springer: Cham, Switzerland, p. 9.

Gupta, R., Garg, V.K. 2008. Stabilization of primary sewage sludge during vermicomposting. *Jour. of Hazardous Materials.* 153(3): 1023-1030.

Ndegwa, P.M., Thompson, S.A., Das, K.C. 2000. Effects of stocking density and feeding rate on vermicomposting of biosolids. *Bioresource Technology*, 71(1): 5-12.

Rigby, D., Caceres, D. 2001, Organic farming and the sustainability of agricultural systems. *Agricultural Systems*, 68(1): 21-40.

9

CHAPTER

Soil Improvement and Amendments

9.1 INTRODUCTION

Organic agriculture refers to 'agriculture without or limited use of chemicals. Soil amendments play a major role in improving the soil structure, texture, quality and health. Soil quality can be improved by using organic fertilizers which are mainly derived from municipal waste and kitchen waste, agriculture waste, etc. Although they contain low concentration of nutrients as compared to synthetic chemicals, but these are multifunctional such as improve soil structure, increase the microbial interaction in soil and soil organic matter, reduce the excess N and P content in water bodies (Antonious 2016). This chapter gives an idea regarding the soil amendments keeping in view the environmental health.

9.2 SOIL AMENDMENT

Soil amendments can be defined as the materials or fertilizers which carry major nutrients for plant growth as well as to condition the soil and to improve its quality. The prime importance of soil amendments is to provide all the necessary nutrients for proper growth of the plant. If the soil amendments are used in excess amount, then it can directly or indirectly affect our environment.

9.2.1 Characteristics of Soil Amendments

- Fast water absorbance capacity
- Natural and earthy.
- High nutrient rich.
- Proper C:N ratio.
- Eco-friendly.
- Slow release.

9.2.2 Functions of Soil Amendments

- Enrich the soil with major and minor nutrients.
- Increase water absorbance capacity and reduce percolation loss.
- Modify the soil structure and physio-chemical properties.
- Make the soil healthy.
- Maintain proper C:N ratio.
- Improve plant growth.
- Increase the microbial interaction in soil.

9.2.3 Importance of Soil Amendments

- Enrich the soil with nutrients which helps in proper growth of the plant.
- Play a major role in improving the soil quality as well as soil health.
- Have the rebuilding capacity of poor soil or the soil which are managed in a improper way.
- Wide range of material can be used as soil amendment.
- Are eco-friendly, locally available, low cost.
- Can be applied to a wide range of plants both in nursery as well as greenhouse condition.
- After a period of time, the soil becomes compact. But these amendments keep the soil healthy and loose.
- Improve the water retention capacity and adjust the pH (Liu et al., 2019).

9.2.4 Types of Soil Amendments

Soil amendments are mainly divided in to 2 groups:

A. Organic soil amendment (Ex: Compost, green manure, crop residue, peat, etc.)

B. Inorganic/synthetic soil amendment (Ex: Synthetic binding agents, mineral conditioners, gypsum, etc.)

9.2.4.1 Organic Soil Amendments

1. **Green manure:** Green manure crops can be directly grown in the soil or can be incorporated with soil to increase the soil nutrient and organic matter content. These crops can be included in crop rotation. The deep roots of GM crop can grow deeply and absorb minerals, nutrients and water from the lower soil zone. These can also activate major potential rhizosphere microorganisms which directly and indirectly help in plant growth. Most leguminous crops are potential GM crops as they can add sufficient amount of nitrogen in the field. It is a yield

booster for the succeeding crop. These can act as cover crop also due to the ability to protect soil from erosion loss and nutrient loss by drainage water. *Crotalaria juncea, Sesbania aculeata,* cow pea, cluster bean is some of the major green manure crops.

Characteristics of GM crops

- Deep root system
- Proper nodular growth
- Profuse foliar growth in early growth stage
- Tolerance to poor soils
- Leguminous plants are preferable

2. **Compost:** The organic matter which is prepared from agriculture residue, kitchen and municipality waste to break down the simpler organic or inorganic matter in a process is known as compost. It can be prepared by adding the waste of kitchen, municipality and agriculture waste in layers in a shaded area for a required period of time. Plastics, polythene cover, heavy metals should be avoided in the mixing pit, otherwise it can lead to unpleasant smell. These are enriched with nutrients and can be applied to all types of plants and fields. It can be used as a fertilizer as well as soil conditioner.

   Application and dose of compost mainly depends on the type of soil, its salt content, and the depth to which it is cultivated. Usually, it is applied on top 6-8-inch soil level and mixed thoroughly. It is advised to go for soil testing before compost application. Ploughing of woody materials bark or wood chips should be avoided in vegetable garden otherwise nitrogen content may be reduced.

   Importance of composting

- It can be prepared from locally available materials but carry the major nutrients in high quantity.
- It carries humic acid which prevents the moisture, nutrient or heavy metals to lose easily from upper soil level.
- Application of compost increases the microbial activity in soil which break down the organic contaminants into the available form.
- It can improve the soil quality as well as soil health.
- It helps in aerobic breakdown of organic material which prevents the greenhouse gas emission.
- It is recyclable and reduces the use of synthetic fertilizers.
- The immediate effect of the compost can bind or degrade the contaminants to mix with soil as a source of nutrients.

3. **Peat moss:** These are partially composted mosses which are in light weight and highly porous and absorb more water. Though it carries very little nutrient content, but very potential to hold the nutrients for longer time period (Snyder et. al., 2016).

4. **Crop residue:** Crop residue refers to the left-over materials after harvesting, training and pruning materials and other farm wastes. Its application can enrich the soil with major nutrients and minerals in farm land (Vargas et. al., 2006). Addition of left-over leaves not only adds nutrients in soil but also protect the plant from pathogens (Clough et. al., 2013). Table 9.1 gives a brief idea about the crop residue and its carrying C:N ratio (approximate).

Table 9.1 Crop residue and its carbon nitrogen ratio

S. No.	Crop residue	C:N ratio
1.	Maize straw	40:1
2.	Blue grass	30:1
3.	Peat moss	58:1
4.	Saw dust hard wood	295:1
5.	Pine	729:1
6.	Alfa alfa	13:1
7.	Small grain straw	80:1
8.	Cattle manure	30:1
9.	Sewage sludge	12:1

5. **Coconut shell mulch:** Coconut shell mulch also plays a major role in amending the soil moisture. It is made up of grinding of coconut shells. The effect remains for a longer period of time as it decomposes very slowly. Its unpleasant smell keeps the harmful organisms or animals away from the plant.

   Some other organic fertilizers like bone meal, blood meal, fish meal, bat and bird droppings can also be used in field to enrich it with minerals and nutrients. Bone meal is a good source of phosphorus which is widely used in field rather than blood meal, bone meal or fish meal as these are costly.

6. **Mushroom compost:** The spent mushroom substrate can be used for compost. Its application requires mixing with wheat straw, corn cobs, peanut meal, lime peat moss, etc. The mushroom grown media can also be used in the field as the mushrooms use very less amount of nutrient. But watering should be done properly before applying it due to the presence of high salt content.

7. **Wood ashes:** These are highly effective in increasing the major soil nutrients. Hard wood ashes are used to increase the soil pH. These are highly effective in increasing the major soil nutrients. Hard wood ashes are used to increase the soil pH and is equivalent to 45% carbonate. Soft wood ashes are less potential as compared to hard wood. The recommended dose for wood ash application rate is 25-50 lb /1000 sq. ft. per year.

9.2.4.2 *Inorganic Soil Amendments*

1. **Synthetic binding agents:** These agents bind particles together to form a stable aggregate. The molecular weight is very high containing long-chain polymers.
2. **Mineral conditioners:** Gypsum is one of the major mineral conditioners mainly for high sodium-rich soils. It is also known as land plaster. It can be used for reclamation purpose for alkali soils and saline soils. It mainly helps to leach the sodium from saline soils. The dose of gypsum application mainly depends on its purity and quantity of sodium present in soil (Thuy et al., 2007).

9.2.5 Sources of Soil Amendment

1. **On farm sources:** It mainly covers compost, crop residue, manure, silage juice, bedding spent soilless media, spent mushroom substrate, etc. Table 9.2 shows the sources of amendments and their use.

Table 9.2 On-farm sources of soil amendments and its use

S. No.	On-farm source of soil amendment	Used as	Remarks
1.	Crop residue	Both fertilizer and soil conditioner	Dry green leaves, stem or pruning Variable nutrient levels and C:N ratios
2.	Compost	Both fertilizer and soil conditioner	Slow nutrient release
3.	Spent mushroom media	Only fertilizer	Variable nutrient levels
4.	Spent nutrient solution	Only fertilizer	Variable or low nutrient levels
5.	Spoiled food	Only conditioner	High BOD, variable or low nutrient levels
6.	Silage effluent	Only fertilizer	High BOD and high nutrients

Cond...

7.	Bedding	Soil conditioner	Can be added with manure Low nutrient content
8.	Solid manure	Both fertilizer and soil conditioner	Can be used as soil conditioner if low nutrient content and C:N ratio is more than 30 : 1
9.	Liquid manure	Only fertilizer	Variable nutrient level
10.	Spent soil-less media	Only soil conditioner	Wood waste or peat based
11.	Milk house waste	Only fertilizer	Can be applied by mixing with liquid manure

2. **Off-farm sources:** Chemical fertilizers or conditioners such as perlite, compost, lime, wood waste and municipal waste are included. Off-farm source soil amendments and their functions are outlined in Table 9.3.

Table 9.3 Off-farm sources of soil amendments and their use

S. No.	Off-farm Source of soil amendment	Used as	Remarks
1.	Commercial fertilizers	Fertilizer	Contain heavy metals and nutrient levels may vary.
2.	Liming substances	Soil conditioner	pH may vary.
3.	Food processing waste	Fertilizer and soil conditioner	Slow release.
4.	Spent mushroom media	Fertilizer and soil conditioner	Nutrient levels may vary.
5.	Off-farm solid manure	Fertilizer and soil conditioner	Can be used as soil conditioner if low nutrient content and C:N ratio is more than 30 : 1.
6.	Whey	Fertilizer	Added with liquid manure.
7.	Fish waste	Fertilizer and soil conditioner	Slow release.
8.	Biosolids	Fertilizer and soil conditioner	Slow release.
9.	Compost	Fertilizer and soil conditioner	Slow release.
10.	Wood waste	Soil conditioner	High C:N ratio

9.3 CONCLUSION

The application of organic soil amendments has several benefits over synthetic chemicals. These soil amendments should not contain any harmful substances which hamper our environment and livestock. The repeated use of these organic products modifies the microclimatic condition of soil and improves the soil physical and chemical structures which help in growth of the plant and boosting the yield. In organic agriculture, inorganic fertilizer can also be used but only the certified one to a limited extent. The use of organic soil amendments and their benefits are still unknown to many rural farmers. So, more emphasis requires on-training and awareness programmes by which will lead to a sustainable organic agriculture in near future.

9.4 QUESTIONS

Long Questions

1. What do you mean by soil amendments? Give a brief idea about this?
2. What is the status of organic soil amendments in India in the context to global view?
3. Why a farmer should choose organic soil amendments over inorganic?
4. What are the benefits of organic fertilizers over inorganic?
5. What is vermicomposting? Discuss the procedure.
6. What is green manuring? What are its benefits?
7. How the plant residues can help to grow the plant better in field? Discuss.

Short Questions

Write Short Notes on:

1. Soil amendments
2. Composting
3. Humus
4. Mushroom spent culture
5. Wood ash
6. Soil conditioners
7. Organic fertilizers
8. Green manuring

9. Sources of soil amendments
10. Coconut shell mulch

REFERENCES

Antonious G.F. 2016. Soil amendments for agriculture production. In Book: Organic fertilizers- from basic concepts to applied outcomes. In Tech open. Chpt 7: 157-187.

Clough TM, Condron LM, Kammann C, Muller C. (2013). A review of biochar and soil nitrogen dynamics. *Agronomy*, 3: 275–293.

Liu ZH, Kanjo Y, Mizutani S. (2009). Removal mechanisms for endocrine disrupting compounds (EDCs) in wastewater treatment—Physical means, biodegradation, and chemical advanced oxidation: A review. *Sci Total Environ*, 407(2): 731-748.

Snyder SA, Wert EC, Rexing DJ, Zegers RE, Drury, DD. (2006). Ozone oxidation of endocrine disruptors and pharmaceuticals in surface water and wastewater. *Ozone Sci Eng*, 28(6): 445-460.

Thuy HTT, Vy NNH, Loan TTC. (2007). Anthropogenic input of selected heavy metals (Cu, Cr, Pb, Zn, and Cd) in the aquatic sediments of Hochiminh City, Vietnam. *Water Air Soil Pollut*, 182: 73-81.

Vargas DN, Bertiller MB, Ares JO, Carrera AL, Sain CL. (2006). Soil C and N dynamics induced by leaf-litter decomposition of shrubs and perennial grasses in the Patagonian Monte. *Soil Biol Biochem*, 38: 2401-2410.

10 Biofertilizers

CHAPTER

10.1 INTRODUCTION

The repeated and long-term cultivation of agricultural lands leads to poor soil condition which lacks several major and minor nutrients. The use of high dose of agro-chemicals pollute the environment in many ways causing a threat to our future upcoming. Therefore, it is essential to choose an eco-friendly pathway to increase the nutrient contents in soil which will unlock the path for sustainable agriculture (Thomas and Singh 2019). The use of biofertilizer not only nourish the soil but also help in increasing yield in a sustainable way. In this chapter, we will study about biofertilizer, its types and its pros and cons which will help us to ensure a qualitative environment for our future generations.

10.2 WHAT IS BIOFERTILIZER?

The ready-to-use formulations, containing living cells or potential strains of micro-organisms which help to increase the nutrient content of soil is known as biofertilizer. Biofertilizers can be applied to seed, root or directly to soil which supports in raising the interaction of microflora near the rhizospheric region. The population of the efficient microorganisms becomes high near plant roots after application of a biofertilizer.

The use of biofertilizer can be considered a major component of integrated nutrient management due to its low cost, high potential and eco-friendly approach. Specific microorganisms are used to enrich the soil with the specific nutrient, e.g., rhizobium increase nitrogen availably as well as Phosphorus, VAM (vesicular-arbuscular mycorrhiza) increase phosphorus, etc (Kumar et al., 2013).

10.3 IMPORTANCE OF BIOFERTILIZER

Biofertilizers are highly beneficial and important in modern agriculture as (Thomas and Singh 2019):

- They maintain the soil fertility and soil health by secreting growth promoting substances.

- They help in increasing the yield.
- These are also known for their ability to suppress the growth of pathogens.
- They improve the soil health as well as soil's physical properties.
- VAM induces drought tolerance in plants by deepening the root, maintaining proper turgor potential, transpiration, etc.
- These are cost effective.
- The eco-friendly and pollution-free character of biofertilizers improves the sustainability in agriculture.

10.4 TYPES OF BIOFERTILIZERS

Biofertilizers can be classified according to the nutrients they enrich or help in uptaking by the plant. These are classified as nitrogen fixing, phosphorus solubilizing/mobilizing, PGPR (plant growth promoting rhizobacteria), etc., which is described in Table 10.1.

Table 10.1 Classification of biofertilizers

S. No.	Groups	Examples
	Nitrogen fixing Biofertilizers	
1.	Symbiotic	*Rhizobium, Anabaena, Azollae, Frankia*
2.	Associative symbiotic	*Azospirillum*
3.	Free living	*Azotobacter, Clostridium, Anabaena, Nostoc*
4.	Endophytic	*Glucanoacetobacter, Burkholdria*
	Phosphorus mobilizing biofertilizers	
1.	Arbuscular mycorrhiza	*Glomus* sp., *Scutellospora* sp.
2.	Ectomycorrhiza	*Boletus* sp., *Amanita* sp., *Laccaria* sp., *Pisolithus* sp.
3.	Ericoid Mycorrhiza	*Pezizella* sp.
	Plant growth promoting Rhizobacteria	
1.	Pseudomonas	*Pseudomonas fluorescence*
	For micronutrients	
1.	Silicate and Zinc solubilizers	*Bacillus* sp.

10.4.1 Nitrogen Fixing Biofertilizers

Microorganisms can fix the nitrogen in three ways, i.e., through symbiotic association, non-symbiotic association and free living. The mechanism is different for all these above-mentioned microorganisms. The free-living bacteria fix the atmospheric nitrogen without forming any association with the plant while symbiotic microorganisms mainly form root nodules and stay in the nodule to fix the nitrogen (Onyia et al., 2020).

10.4.1.1 Rhizobium

Family- Rhizobiaceae

Genus: *Rhizobium, Azorhizobium, Bradyrhizobium, Sinorhizobium*

Rhizobium is one of the most important symbiotic nitrogen fixing bacteria. This bacterium was first discovered by Beijirinck from root nodules in 1888 and named as *Bacillus radicola*. The name *Rhizobium* was given by Frank in 1889. These bacteria can live freely in soil but undergo to symbiotic association only with the leguminous plants. They enter the root through root hair, secrete stimulatory exudates and form nodule there. They facilitate the nitrogen uptake for the plant while in exchange, they get shelter in the root from the host plant. The nodules contain leghemoglobin pigment which is pink in colour and termed effective nodules. This pigment helps in nitrogen supply and support the activity of nitrogenase enzyme. This nitrogenase enzyme plays a major role in reduction of nitrogen to ammonia in N-fixation.

When the leguminous plants form root nodules after inoculating rhizobium obtained from any other legume plants, it is known as cross inoculating group (Table 10.2). The rhizobium is classified as slow growing rhizobia, e.g., bradyrhizobia and fast-growing rhizobia, e.g., rhizobium.

Table 10.2 Cross inoculating groups

S. No.	Rhizobium species	Principal host
1.	*Rhizobium loti*	Lotus
2.	*Rhizobium meliloti*	Medicago (Lucerne), Trigonella (fenugreek)
3.	*Rhizobium etli*	*Phaseolus vulgaris, Mimosa affinis*
4.	*Rhizobium hainanense*	Sesbania
5.	*Bradyrhizobium japonicum, B. elkanii*	Glycine
6.	*Bradyrhizobium* spp.	Vigna, Cicer, Lupinus
7.	*Bradyrhizobium yuanmingense*	*Lespedeza, Medicago, Melilotus*
8.	*Azorhizobium caulinodans*	*Sesbania rostrata*
9.	*Sinorhizobium indiaense*	*Sesbania rostrate*
10.	*Biovar viceae*	*Vicea*
11.	*Biovar trifolii*	*Trifolium (Berseem)*
12.	*Mesorhizobium ciceri*	*Cicer arietinum*

The amount of nitrogen fixed varies according to the strains of rhizobium, abiotic factors and host plant species which is described in Table 10.3. Seed treatment is the most appropriate method for rhizobium inoculation.

Table 10.3 Amount of nitrogen fixed in different legumes and non-legumes by symbiotic microorganisms

S. No.	Plants	Kg/ha
	Leguminous plants	
1.	Cowpea	73-354
2.	Soybean	1-168
3.	Green gram	61-342
4.	Chickpea	85-100
5.	Lupin	145-208
6.	Pigeon pea	168-280
7.	Lucerne	184-463
	Non-Leguminous plants	
8.	Casuarina	52
9.	Alnus	139

10.4.1.2 Blue-Green Algae (BGA)

Blue-green algae (BGA) are also known as cyanobacteria, cyanophytes or cyanoprokaryotes. These are very small and can be seen under microscope. But some BGA accumulate in large quantities and can easily be distinguished. They are similar to algae to some extent. BGA carry a specific type of gas vesicle in their cells which facilitates floating on water bodies and goes to the bottom of water depending their light and nutrient availability.

BGA (*Anabaena azollae*) live in the dorsal side of the aquatic fern Azolla in symbiotic relationship and help in fixing atmospheric nitrogen. Both BGA and Azolla are suitable in rice fields and increase the uptake of nitrogen by the paddy (Simarmata et al., 2016). Azolla releases nitrogen slowly but it does not leach out. It can also be used as organic manure as well as a major protein-rich food for fish and poultry. In addition to adding nitrogen in rice fields, BGA also release growth promoting substances like auxin and amino compounds which facilitate plant growth. Algae can be applied as broadcasting in the rice field at 10 kg/ha after 1 week of transplanting.

10.4.1.3 *Azospirillum*

Azospirillum was first time isolated by Beijerinck from the roots of *Paspalum* in 1922. In 1978, Tarrand and his co-workers named it as *Azotobacter paspali*. Later, the name changed to *Spirillum lipoferum*.

It is a Gram-negative, aerobic free living and associative symbiotic bacterium. This bacterium fixes nitrogen mostly in C_4 plants such as maize, sugarcane, sorghum, etc., but also found to be close association with higher plants. These bacteria stay on plant roots and sometimes invade the plant roots. They fix nitrogen up to 10-40 kg/ha without producing any nodule or out growth. Additionally, they also secrete plant growth promoters which develop the root hair, suppress the growth of other harmful microorganisms, facilitate water and other mineral uptake and help in reducing the toxicity of compost. It is cost effective as it reduces the use of nitrogenous fertilizer up to 25 to 30%. Among the identified four species (*A. lipoferum, A. brasilense, A. amazonense and A. iraquense*), only two species (*A. brasilense and A. lipoferum*) are found in India (Abd El-Lattief. 2016).

10.4.1.4 *Azotobactor*

It is a heterotrophic, free living, aerobic, Gram-negative and N-fixing bacterium and mostly found near the rhizosphere region (Bhardwaj et al., 2014). The major characteristic feature to identify these bacteria from other species are cell size, pigmentation, flagellar arrangements and extracellular slimes. In India, *A. chrococcum* is the most common species. They can produce cyst and secrete gum as well as poly beta hydroxy butyric acid. It can grow in alkali soils and organic matter-rich soils. It can fix 20-40 kg nitrogen per hectare. It reduces the use of nitrogenous fertilizer up to 20%. It secretes growth promoting substances like GA, IAA, etc. Additionally, they are capable of inhibiting the growth of *Fusarium, Aspergillus*. It improves the plant growth and seed germination. It is mainly applied in rice, wheat, millets, vegetables, some plantation crops and flowers.

10.4.1.5 Acetobactor

Acetobactor diazotrophicus, an endophytic N-fixer is mainly found in association with the leaf, stem and roots of sugarcane crop. Cavalcanti and Dobereiner first isolated this bacterium in 1988 and named it *Gluconacetobacter diazotrophicus* which later on named *Acetobactor diazotrophicus.* It can also be found in the fluid of apoplastic region and xylem vessels of sugarcane plant. It can tolerate high sucrose and salt concentration. It is highly beneficial for sugar crops growing in acidic soil. These are capable of fixing 200 kg of nitrogen per hectare. In addition to this, these can secrete plant growth hormones like auxin, GA and also antibiotics.

10.4.1.6 *Frankia*

An actinomycetes, *Frankia* can stay in symbiotic association with *Alnus* and *Casuarina* plants. This is also capable of forming root nodules in this leguminous plants and fixes atmospheric nitrogen. It can also help in plant growth as well as increase soil fertility in wastelands. This is one of the major groups of microorganisms which plays an important role in increasing the nutrient uptake in nitrogen deficient agro-forestry system.

10.4.2 Phosphorus Solubilizing Microorganisms (PSM)

The heterotrophic microorganisms which make the environmental phosphorus available for the plant is known as phosphorus solubilizing microorganisms. Various bacteria and fungi are capable of solubilizing the insoluble phosphorus. But the population and role of fungi is more in phosphorus solubilization. These are able to secrete organic acids, enzymes and mineralize organic form of phosphorus. They are also capable of producing citrate and water-soluble phosphorus when applied to rock phosphate, bone meal or slag. These can induce the composting process and can suppress the growth of competitive pathogens. The major examples of PSM are *Bacillus, Aspergillus, Trichoderma* and *Penicillium*.

10.4.2.1 *Vesicular Arbuscular Mycorrhiza (VAM)*

The name 'mycorrhiza' was first time coined by A.B. Frank in 1885. This word is taken from a Greek word, i.e., 'mukes' means fungus and 'rhiza' means root, thus 'fungus root'. It is symbiotically associated with plant roots. It lives both inside and outside of the plant root. They facilitate the phosphorus, zinc, sulphur uptake of plant as well as transfer of iron, manganese, aluminium from soil to plant root. It also helps in water uptake, root growth, pathogen growth suppression by competition, increase plant tolerance against some abiotic stresses, etc.

The mycorrhiza which lives inside the roots is known as endomycorrhiza, living outer surface of root is known as ectomycorrhiza and which lives both outside and inside is known as ectoendomycorrhiza. The ectomycorrhiza grows outer surface of root, covers the root by mycelium which is termed Hartig net, a site for nutrient exchange (Ansori and Gholami 2015).

Mycorrhiza enters the root cortex and then moves forwards intracellularly. They carry a sac-like structure to store nutrients which is referred to as vesicle. Arbuscular is another structure which helps them to penetrate the root system. Difference ectomycorrhiza and endomycorrhiza are shown in Table 10.4.

Table 10.4 Comparison between endomycorrhiza and ectomycorrhiza

Characters	Endomycorrhiza	Ectomycorrhiza
Association	Agricultural, horticultural and tropical trees	Forest trees
Structure	Vesicles and arbuscules inside the root	Majority of the structure outside of the root system
Branching of hypha	Loose networking in soil and extensive growth in roots	Branching outside of the root
Changes	No morphological changes in roots	Morphological changes in root
Culture	Can't be cultured on artificial media	Cultured on artificial media
Example	Glomus, Gigaspora, Sclerocystis, Endogone, Acaulospora, etc.	Amanita, Entomoloma, Boletus, Rhizopogon, etc.

10.5 METHODS OF BIOFERTILIZER APPLICATION

Application of biofertilizers mainly depends on the type of seeds and type of biofertilizer (Thomas and Singh 2019). It can be directly applied in soil, seed or seedling. The various methods of applications are described below.

10.5.1 Soil Inoculation

Soil inoculation of biofertilizers is mainly suitable for plants where localized application is required like fruit trees. 20 g of biofertilizer mixed with compost is given in the ring of one plant at the time of planting. After the plant matures, again the same quantity of biofertilizer is given. Broadcasting of biofertilizer is not advisable as it require more amount.

10.5.2 Root and Seedling Treatment

This method is mainly suitable for Cole crops, tomatoes, onion, rice and flowering plants. In this method, the plants are dipped in the biofertilizer solution for 30 min before transplanting. For one acre area, 2-2.5 kg biofertilizer is mixed with water properly in a drum or bucket and then the ready to be transplanted plants are dipped. The plants will absorb this solution which will help them to get more nutrient from soil.

10.5.3 Seed Inoculation

Rhizobium is mainly used for seed treatment in leguminous plants (Cortivo et al., 2020). A slurry solution is prepared by using jaggery which acts as a proper sticker among seeds and biofertilizer. For seeds of one hectare of land 900 g of soil base culture is required. Jaggery is initially mixed thoroughly with the seeds and after that it is be spread to be dried. After getting a proper coating on seed, the rhizobium culture is mixed properly and dried under shade for overnight. It is advised to sow the treated seeds in the evening for better performance.

10.5.4 Self-Inoculation

This method is mainly applicable for azotobacterial based biofertilizer. It is advisable to treat any planting material like potato through this method. In this method, 4-5 kg of biofertilizer is mixed with 50 litres of water. The planting materials are dipped in this solution and then kept in shade for drying. After drying, they can be planted in the evening.

10.6 ADVANTAGES

- Increase the nutrient availability in soil as well as the uptake capacity of plant
- Secretion of plant growth regulators and hormones helps in the growth of plants.
- Release of antibiotics suppresses the growth of harmful microorganisms.
- Increase soil fertility and soil health for a longer time.
- Cheap and eco-friendly in nature.
- Enrich the soil with other minerals and nutrients.
- Increase the yield and returns more profit.

10.7 DISADVANTAGES

- Cannot be stored for long time.
- When applied in field, their potential may decrease due to biotic and abiotic stresses. Some biofertilizers depend on the soil type like acidic, alkaline or salty soil.
- Are specific to the plants.
- Are slow responders as compared to conventional biofertilizers.

10.8 LIMITATIONS

- Are yet to be popular among farmers as most farmers are not aware of it.
- Require more promotion and sufficient awareness programmes.
- Require high skill for application as well as production.
- Due to various abiotic stresses, their quality decreases after application.

10.9 CONCLUSION

Biofertilizers are one of the major items for organic agriculture and sustainable agriculture. They not only increase the soil nutrient level but also promote growth, increase yield, defend the plant from pathogens, improve soil health, etc. Microorganisms such as *Rhizobium, Frankia, Azotobacter, Azospirillum, Acetobacter,* VAM, etc., play a crucial role in nutrient enrichment in soil. They can show their potential effect in a wide range of plants, various geographic areas and different types of soil. Their multiple potential effects have made biofertilizers more demanding in recent days to build up for a better environment through sustainable farming.

10.10 QUESTIONS

Long Questions

1. Define biofertilizer. Chalk out their advantages, disadvantages and constraints in their application.
2. Explain the various methods of biofertilizer application.
3. Define the term 'mycorrhiza'. Briefly explain VAM.
4. Why a farmer should choose a biofertilizer in place of any other conventional fertilizer? Explain.
5. List out all N-fixing microorganisms. Explain briefly about all the microorganisms.

Short Questions

1. What do you mean by biofertilizer?
2. How rhizobium can be applied to increase nitrogen uptake?
3. What is the difference between ectomycorrhiza and endomycorrhiza?
4. What are the constraints in biofertilizer application?
5. Write the method of biofertilizer application in case of seedlings.
6. List out the microorganisms used in biofertilizer for various nutrient uptakes.
7. What do you mean by cross inoculating group?
8. What do you understand by VAM?

REFERENCES

Ansori A, Gholami A. (2015). Improved nutrient uptake and growth of maize in response to inoculation with *Thiobacillus* and mycorrhiza on an alkaline soil. *Commun Soil Sci Plant Anal*, 46: 2111-2126.

Bhardwaj D., Ansari M.W., Sahoo R.K., Tuteja N. (2014). Biofertilizers function as key player in sustainable agriculture by improving soil fertility, plant tolerance and crop productivity. *Microbial Cell Factories*, 13(1): 1-10.

Cortivo C.D, Ferrari M. et al. (2020). Effects of Seed Applied Biofertilizers on rhizosphere biodiversity and growth of common wheat (*Triticum aestivum* L.) in the field. *Front. Plant. Sci*, 11: 72.

Kumar K., Shukla N. et al. (2013). Biofertilizers for organic agriculture. *Popular Kheti*. 1(4): 91-96.

Onyia C.O., Okoh A.M. and Irene O. (2006). Production of plant growth promoting bacteria biofertilizer from organic waste material and evaluation of its performance on the growth of corn. *American Journal of Plant Science*, 11(2): 1-12.

Simarmata T., Hersanti et al. (2016). Application of bio ameliorant and biofertilizers to increase the soil health and rice productivity. *HAYATI journal of Bioscience*s, 23(4): 181-184.

Thomas L. and Singh I. (2019). Microbial biofertilizers: Types and applications. In: *Biofertilizers for Sustainable Agriculture and Environment*. Springer Nature Switzerland.

Abd El-Lattief E.A. (2016). Use of *Azospirillum* and *Azobacter* bacteria as biofertilizers in cereal crops: A review. *IJREAS*, 6: 36-44.

Unit 4

Operational Structure and Biotic Stress Management

CHAPTER 11

Integrated Diseases and Pest Management: Use of Biocontrol Agents, Biopesticides, Pheromones, Trap Crops, Bird Perches

11.1 INTRODUCTION

Plant disease is deterioration of the normal state of a plant that interrupts or modifies its vital functions which leads to compelling crop losses. Integrated disease management (IDM) is the cumulative approach of all the suggestive measures including biological, cultural, physical and chemical control strategies. The main objective of IDM is to protect crops by feasible eco-friendly measures, for better surroundings in addition to sustainable improvement. Different kinds of microorganisms like fungi, bacteria, viruses, phytoplasmas, and nematodes act as a major biotic constraint for disease development and its spread. By bringing some new agenda or recent policies like advancement in plant protection, investment in the private sector, trade and export, awareness and education, food safety, land prioritization and agricultural extension policies. Successful adaptation and implementation of IDM strategies is the prerequisite agenda in developing countries. By crossing every barrier and technical competencies from the farmers level makes IDM more enriched with insightful knowledge of field management, crop rotation, productivity improvement as well as cost effectiveness.

11.2 PRINCIPLES OF IDM

The term 'control' denotes for the permanent solution whereas 'management' signifies a continuous process to make a barrier in adverse conditions. In broader spectrum organic farming includes diverse aspects of soil health management, plant health management, animal health management, post-harvest management, environmental management, water quality management, etc. (Figure 11.1) whereas disease management requires a detailed understanding of all aspects of crop production, economics, environmental cultures, genetics and epidemiological information. There are six different principles of IDM.

11.2.1 Avoidance of the Pathogen

Before introducing a pathogen/inoculum in a particular area, disease can be avoided by alteration of planting or sowing time or when inoculums remain inactive due to environmental conditions. This principle follows the methods of choice of geographical area, selection of a field, adjustment of time of sowing, use of disease escaping varieties, use of pathogen-free and planting material and modification of cultural practices.

11.2.2 Exclusion of the Pathogen

This strategy can be followed by preventing the inoculums from entering or establishing in a field or a geographical area by applying legislative measures like quarantine regulations are needed to be strictly followed to prevent disease spreading. This principle is fulfilled by applying the methods like treatment of seed and plating materials, inspection and certifications, eradication of insect vector, quarantine regulations.

11.2.3 Eradication of the Pathogen

This process includes inactivating, reducing, eliminating or destroying inoculums at the source, either from the specific region or an individual plant (roughing). These principles are fulfilled by application of biological control agents, eradication through alternate and collateral hosts, cultural methods (like crop rotation, sanitation of field, removal and destruction of diseased plants, roughing), heat and chemical treatment of disease plants, soil treatment (heat energy, chemicals, flooding and fallowing), etc.

11.2.4 Protection of the Host

Host plant can be protected from the disease or pathogens by applying some chemicals or bioactive agents by which disease can be controlled. The aforesaid principle includes application of chemicals (antibiotics, fungicides) by seed treatment, dusting and spraying, modifications of environment, modifications of host nutrition, etc.

11.2.5 Disease Resistance

In this strategy host plants, preventing infection or reducing the effect of the infection of the pathogen through auto-resistance (as resistance variety through genetic manipulation) or adoptive resistance (by chemotherapy). Under this principle host nutrition and genetic engineering, tissue culture is also considered.

11.2.6 Application of Therapy

In this strategy infected individual is treated by different types of therapy (chemotherapy, heat therapy, tree surgery) to overcome the pathogenic toxin and hazards. All the five above-mentioned principles are prophylactic (preventive) whereas this principle is curative.

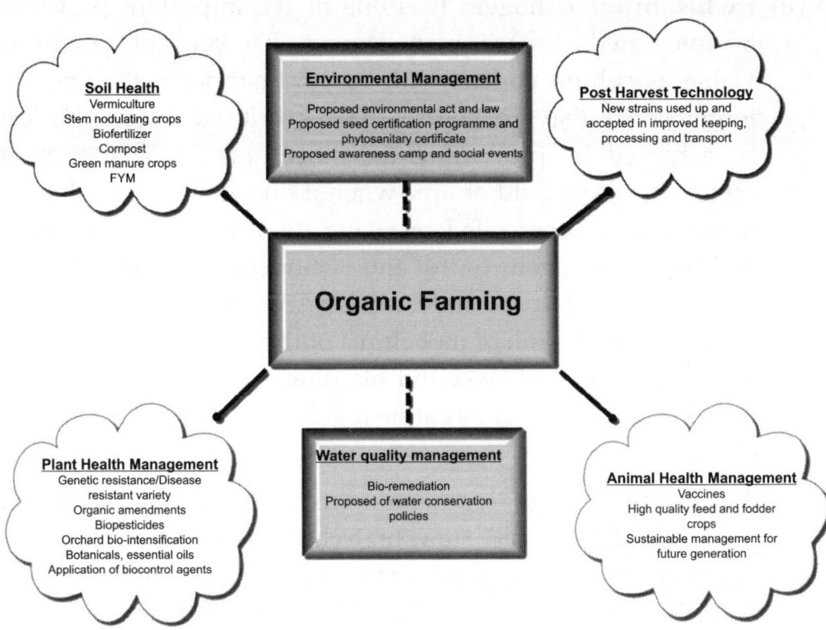

Fig. 11.1 Organic farming interlinkage with various sectors and sub-sectors.

11.2.7 Disease Management Strategies in Organic Farming

In traditional and modern agricultural systems we use different strategies to make IDM stronger. Some of the strategies described below.

(i) **Growing disease-resistant varieties:** Resistant crop cultivation is one of the major and foremost important tools under integrated disease management. Implementation of low-cost cultivation is successful in terms of using resistant varieties, where other farming expenditure is near about zero. Among all the strategies, introducing resistant varieties by genetic modification, gene editing and plant breeding (pure line and mass selection) to resist the pathogen interaction with the host is one of the best methods. According to gene-for-gene hypothesis, one gene of the host and one gene of the pathogen must be compatible for successful disease development. But in case of resistant varieties, the disease escapes and the growth of pathogen can be reduced through modification in their maturity period, nutrient content, growth habit, plant architecture, etc. Care should be taken for selection of resistant varities in terms of geographical compatibility and dominance of a specific pathogen in a particular region as the resistant varieties are very specific to particular diseases.

(ii) **Exclusion of pathogen:** It is one of the important principles of integrated disease management. To prevent potent or viable inoculums contact with the host surface is the first and foremost strategy to control the severity disease. This can be achieved by preventing the inoculums from entering or establishing in a field or area when it does not exist. Soil-borne pathogens are controlled by using disease-resistant varieties. Soil-borne pathogen control and vector control can be done by using anaerobic soil disinfestation (ASD) and soil solarization technique. Exclusion of inoculums of the pathogen can be done by: (a) treatment of seed and planting material, (b) inspection and certification, (c) quarantine regulations, and (d) eradication of insect vectors.

(iii) **Application of organic amendments:** Mostly soil enriched with low microbial diversity show high chance of infection. Good soil health condition is a key factor to suppress different weed, pests and diseases. Increasing soil health condition by incorporating green manures, cover crops, animal manures, fermentated and decayed plant debris are used for long-term solution. By this technique not only restricting plant pathogenic microbes but also increasing biological activity of soil through enrichment of different microbial populations, which are also known as the plant growth promoters. Application of composts and green manures increases soil microbial diversity which consecutively helps in pathogenic suppressiveness. Organic amendments are mostly biodegradable and also easily available. Generally, Neem cakes are used for soil amendments to control nematodes and soil-borne pathogens at the rate of 0.25-0.5 f/ha.

(iv) **Cultural control:** Cultural control strategy strictly indicates good agricultural practice by which it can promote good soil health and good plant health. Alteration or shifting of planting schedule, field sanitation, weed management decreases disease incidence as well as pathogenic spread. Crop rotation is also an effective process to control pathogenic spread, weed and disease control. As par Litterick et al. (2002) in organic farming pest or disease control strategy is preventive rather than curative. By alternate cropping, mixed cropping, inter- and intra-cropping patterns by giving different choices to virulent pathogens every time we try to provide a barrier by which virulent pathogen and susceptible host do not come in contact easily. Though in every case it has a positive side and a negative side. Such as crop rotation is not effective if the pathogen shows long survivality under the soil. Poor irrigation system in the field not only toxifies the soil but also brings a favourable condition for plant pathogenic microbial populations. It is also found some pathogens survive in dead and decayed plant debris. Proper tilling and cleaning of the plant debris, number of potential pathogen populations reduces.

(v) **Orchard bio-intensification:** Orchard bio-intensification is a strategy by which habitat modification of beneficial microorganisms is maintained, such as development of healthy and biologically fertile and active soils, maintaining uncultivated lands for diversity of flora and fauna, development of entomophage parks within orchard for food and shelter to diverse beneficial insects, weed strips, wind breaks, hedge rows, stagnant water bodies, inter-crops and conservation of insect biodiversity, etc.

(vi) **Physical methods:** One of the important processes is soil solarization. Soil solarization helps to reduce the pathogen population at seed bed or nursery bed level. Hot water treatment or stream treatment is widely used for seeds/planting material treatment. It is also observed that post-harvest hot water treatment reduces the anthracnose and different diseases.

(vii) **Botanicals, essential oils:** Use of different types of botanicals from prehistoric era in traditional agriculture is a common practice. Cow urine, neem oil are also used to control various diseases and insect pests. Application of plant based essential oils is another type to fight against different types of phytopathogens. Hot weather condition is considered to be a natural epidemiological barrier. Treatment by coupling with neem oil and soft soap solution are very much effective against vectors like aphids. Essential oils are considered to be volatile, natural, complex compounds from different plant families (Asteraceae, Myrtaceae, Lauraceae, Lamiaceae). Chemical compounds have the property of insecticidal, acaricidal, antibiotic and can control diverse plant pathogens. The volatile substance contains tarpenes, terpenoids and different allopathic and aromatic components. Different types of botanicals classified based on the plant parts used, bioactive components and property are represented through the following five tables (Tables 11.1 to 11.5)

Table 11.1 List of botanicals used in organic farming

Sl. No.	Common name	Scientific name	Active chemical constituents	Property
1.	Tulsi	*Ocimum sanctum*	Linalool, chavicol, eugenol, eugenol methyl ether, cineole, caryophyllene.	Insecticidal repellent
2.	Akanda	*Calotropis procera*	Latex containing poisonous constituents	Antifeedant
3.	Neem	*Azadirachta indica*	Limonoids, azadirechtins, salamin, nimbin	Insecticidal, hormonal (JH), antifeedant, multifacial
4.	Kalmegh	*Andrographis paniculata*	Andrographolide	Insecticidal
5.	Basak, Adusa	*Adhatoda vasica*	Vasicine, vasicinone, vasicinol, limonene	Insecticidal, antifeedant
6.	Nigundi	*Vitex negundo*	Rotundial	Repellent, insecticidal
7.	Mehandi	*Lawsonia inermis*	Tannin, saponin, anthraquinone flavonoids, glucosides and alkaloids	Antifeedant
8.	Dill	*Anethum sowa*	Carvone, dillapiole	Insecticidal
9.	Safeda	*Eucalyptus hybrid*	1,8-Cineole, α-phellandrene, linayl isovalerate, isoamyl isovalerate, etc.	Antifeedant
10.	Satyanashi, Sialkata	*Argemone maxicana*	Protopine nitrate, Berberine nitrate, Ceryl alcohol, β-sitostero	Protection
11.	Raimuniya	*Lantana camera*	Cayophyllene, cineol and β-pinene	Protectant
12.	Seeta-bani	*Artemisia capillaris*	Bornyl acetate, capillarin, capillen	Feeding deterrent
13.	Nimbu	*Citrus limon*	Limonin, nomilin, obacunone	Antifeedant, toxicant
14.	Bachh	*Acorus calamus*	*Trans*-asarone, *cis*-asarone, isoasarone	Repellent, antifeedant
15.	Camellia	*Camellia* spp.	Shikinic acid, caeffin and tannins	Insecticidal repellent
16.	Papaya	*Carica papaya*	Carpaine	Insecticidal
17.	Jangli Tamatar	*Lycopersicon hirsutum*	2-tridecanone, *trans*-caryophyllene	Repellency, toxicity

18.	Chakunda, Kasonda	*Cassia occidentalis*	Emodin	Insecticidal
19.	Charota	*Cassia tora*	Chrysophanic-9-anthrone	Antifeedant
20.	Balsam fir	*Abies balsamea*	Juvabione, dehyojuvabione	Hormonal (JH)
21.	Nimbu	*Citrus* spp.	Citropin, di-limonens, linalool, glucosides, terpenes, etc.	Insecticidal
22.	Nimbu ghas	*Cymbopogan* spp.	γ-cardiaene, elemicin, citral	Insecticidal repellent
23.	Datura	*Datura metel*	Hyoscine	Antifeedant
24.	Ramphal	*Annona reticulata*	Anonaine, liriodenine, reticuline, norushinsunine	Insecticidal
25.	Amrood	*Psidium guajava*	β-sitosterol, maslinic acid, guijavalic acid	Insecticidal, repellent
26.	Bhang	*Cannabis sativa*	Resinoid, tetrahydrocannabinol, a phenolic type substance	Protectant
27.	Balkuwari	*Ginkgo biloba*	Salicyclic acid derivatives, bilobalide, ginkgolide-A and B	Feeding detterent
28.	Kaner	*Nerium oleander*	Cardiotonic, oleandrin, neridin	Inhibit oviposition
29.	Behaya	*Ipomea carnea*	Alantolactone	Insecticidal
30.	Ratanjot	*Jatropha carcus*	Isovitexin, vitexin, β-sitosterol, Curcine, curcasin, fatty acids, etc.	Protectant, repellent
31.	Murray red gum	*Eucalyptus rostrata*	1,8-Cineole, ∞-phellandrene, etc.	Anti-fecundity
32.	Pyaj	*Allium cepa*	Quercetin and phenolic compounds	Insecticidal
33.	Bakain	*Melia azedarach*	Tatra tarpenoids, toosendanin, meliandiol, melianone, meliantriol, nimbolidin A, volkensin	Antifeedant, oviposition, deterrent, antifertility, toxicant
34.	Senjana	*Moringa oleifera*	Niazirin, niazirinin	Growth inhibitor
35.	Bael/Bilva	*Aegle marmelos*	Limonene, p-caryophyllene, sabinene, ocimene	Fungicidal, feeding deterrence

36.	Moti Saunf	*Foeniculum vulgare*	Fenicularin	Repellent
37.	Soybean	*Glycine max*	Glyceollins, daidzein	Antifeedant, toxicant
38.	Karanj	*Pongamia pinnata*	Karanjin	Apecidal, insecticidal
39.	Cassia	*Cassia nigricans*	Emodin	Insecticidal
40.	Ram Tulsi	*Ocimum basillicum*	Juvocimene I, II, linalool, methyl chavicol, eugenol	Antifeedant, toxicant

Table 11.2 Shoot, flower, fruits and seeds used as botanical biopesticides

Sl. No.	Common name	Scientific Name	Plant parts used	Active chemical constituents	Property
1.	Tamak	*Nicotiana tabacum*	Seeds	Nicotine, oleandrin, neridin	Inhibit oviposition
2.	Latakaranja	*Caesalpinia crista*	Seeds	Karajin, fatty acids	Antifeedant, insecticidal repellent
3.	Ramphal	*Annona squamosa*	Seeds	Anonaine, liriodenine, reticuline, norushinsunine	Insecticidal
4.	Kapur	*Cinnamomum camphora*	Shoot	Camphor oil	Insecticidal
5.	Bamboo	*Bambusa arundinacea*	Fresh and young shoots	Benzoic acid, cyanogenic glucoside	Insecticidal
6.	Palash	*Butea monosperma*	Flowers	Chalcones and Aurones	Termicide
7.	Latkan, Annatto	*Bixa orellana*	Seed coats	Bixin	Repellent
8.	Dill	*Anethum sowa*	Seeds	Cavone, dillapiole	Insecticidal
9.	Sharifa	*Annona reticulata*	Fruit and seed extracts.	Annonacin, annonin, annonelliptine, asimicin, annonidines	Antifeedant, repellent
10.	Genda	*Tagetes minuta*	Flowers	Terthienyl-2,2 ,5, 2terthiophene), E-ocimenone	Larvicidal, repellent

11.	Sarisha	*Brassica comprastis*	Seeds	2-Phenylethyl isothiocyanate	Fecundity reducing
12.	Yellow prickle	*Zanthoxylum monophyllum*	Fruits	1,8 cineole, *trans*-sabinene, hydrate and *cis*-sabinene hydrate	Insecticidal
13.	Chenopo-dium	*Chenopodium anthelminticum*	Seeds	Essential oil having ascaridole	Insecticidal
14.	Lal mirch	*Capsicum frutescens*	Fruits	Capsaicin	Insecticidal
15.	Pudina	*Mentha oleifera*	Flowering tops	Menthol, caryophyllene, carvone, Cineole	Antifeedant, toxicant
16.	Kaju	*Anacardium occidentale*	Cashew-nut shell oil	Phenolic constituents	Insecticidal
17.	Dadmurdan	*Cassia alata*	Seeds	Chrysophanol, glycosides, aloeemodin, Cassiaxanthone, isochrysophanol, β-sitosterol rhein physicion-1-glucoside	Meamorpho-sis inhibitor
18.	Ritha	*Sapindus mukorossi*	Seeds	Saponins	Insecticidal
19.	Sharpunkha	*Tephrosia purpurea*	Seeds	Ratenoids	Insecticidal
20.	Calmogra, Jangli badam	*Hydrocarpus* spp.	Seeds	Hydrocarpic acid, chaulmoogric acid, gallic acid and other fatty acids	Repellent

Table 11.3 Roots/rhizomes/bark/bulb used as botanical biopesticides

Sl. No.	Common name	Scientific name	Parts used	Active chemical constituents	Property
1.	Yellow prickle	*Zanthoxylum monophyllum*	Bark	Zanthophylline	Feeding deterrent
2.	Pyaj	*Allium cepa*	Bulbs	Quercetin and phenolic compounds	Insecticidal
3.	Agarwood	*Aquilaria malacensis*	Agarwood dust	A-guaiene, caryophellene oxide, eudesmol	Protection, repellent

Contd...

4.	Haldi	*Curcuma longa*	Rhizome	α- phellandrene, curcumene, termerone, dehydro-termerone	Protectant, repellent
5.	Rasun	*Allium sativum*	Bulbs	Diallyl di-sulphide, diallyltri-sulfide	Insecticidal
6.	Khas	*Vetiveria zizanioides*	Roots	Β-vetivene, azulene, zizanene, leavojujenol	Growth disrupter, repellent
7.	Safed til	*Sesamum indicum*	Roots	Sesamolin, Sesamin, sesangolin	Antifeedant
8.	Derris	*Derris elliptica*	Roots	Rotenone and dihydrorotenone	Insecticidal
9.	Adrak	*Zanthoxylum monophyllum*	Rhizomes	Gingerdione, paradol	Antifeedant
10.	Cinchona	*Cinchona officinalis*	Bark	Quinine, quinidine, cinchonine & cinchonidine	Insecticidal

Table 11.4 Secondary metabolites, latex used as botanical biopesticides

Sl. No.	Common name	Scientific name	Active chemical constituents	Property
1.	Murry red gum	*Eucalyptus rostrata*	1, 8-Cineole, α-phellandrene	Anti-fecundity
2.	Tridhara	*Euphorbia antiquorum*	Latex contents 4.0-6.4% caoutchouc	Antifeedant

Table 11.5 Whole plant used as botanical biopesticides

Sl. No.	Common name	Scientific name	Active chemical constituents	Property
1.	Indian Aconite, Bishnag	*Aconitum ferox*	Psuedaconitine, diacetyl pseudoaconitine, chasmaconitine, indaconitine, bikhaconitine	Aphicidal toxic to beetles
2.	Nimbu	*Citrus* spp.	Linalool, glucosides, terpenes, citropin, etc.	Insecticidal

(viii) Baking soda, buttermilk: Researchers revealed that buttermilk spray is an effective management strategy against blight, mildew, mosaic viruses and other fungal and viral diseases. Fermented butter milk is used to control severe diseases and pests. Fermented cow dung used as organic manure is a rich source of cellulose degrading, P-solubalizing, N_2-fixing, plant growth promoting microbial populations. Application of baking soda is also effective against mildew and rust diseases of plants.

(ix) Application of biocontrol agents: Microbial biocontrol agents are isolated from natural environmental conditions and used against different kinds of diseases and pests without disturbing the host or toxification of the environment. These viable biocontrol agents are long lasting, ecofriendly and can be widely used for sustainable agriculture on long-term basis. The mechanism of action shows competition, enhanced nutrient uptake, antibiosis, antagonism, induction of host resistance, etc. Unlike chemical pesticides they ate non-toxic for human beings and other non-targeted organisms. Commercially these are available in powder, granule or liquid form. Some of the important biocontrol agents are *Pseudomonas* spp., *Bacillus* spp., *Trochoderma* spp., etc. Biocontrol agents also show plant growth promoting activity and play an important role as growth enhancer.

(x) Application of mineral based fungicides: Researches proved that prophylactic spray of sulphur dust is very effective against downy mildew, powdery mildew and several other diseases. The mechanism involved blocks their spore germination. Bordeaux mixture is an important fungicide (acts against anthracnose, downy mildew, powdery mildew, leaf spots, etc.) and shows both fungicidal and bactericidal properties. Copper and sulphur based fungicides are also used in this process. Copper based products like copper hydroxide, copper oxychloride are applied in organic farming to maintain and moderate copper accumulation in the soil.

(xi) Biopesticides: Biopesticides are those microorganisms or plant products that have the potential to dominate an another kind of pathogenic microorganisms. Generally, in this category some specific microorganisms belong to fungi, bacteria or viruses. Origin of biopesticides may be of plant origin or comes as dominant microorganisms. Biopesticides are not harmful except to targeted hosts and are eco-friendly. The bioactive components in plants are alkaloids, phenol, sterols, quinines, alcohols, tarpenes, saponins, etc.

(a) Advantages of biopesticides

- Are target specific in nature and non-hazardous for beneficial microorganisms.
- Are eco-friendly and help in long-term sustainable agriculture.
- Are biodegradable in nature and do not affect any of the edaphic factors, water or soil.
- Can wipe out very low amount of pollutants through elimination mechanism.
- Required in very low quantity, show highest results and are less toxic.
- Insect pests or targeted pathogen cannot show any resistant mechanism easily, because it's a natural enemy.
- Are non-hazardous, non-toxic as compared to inorganic pesticides or fungicides.

(b) Pesticides registered under Insecticides Act (1968) in Indian context

- *Trichoderma viridae*
- *Trichoderma harzianum*
- *Pseudomonas fluoresens*
- *Beauveria basssiana*
- *Metarhizium anisopliae*
- *Bacillus thuringiensis var. israelensis*
- *Bacillus sphaericus*
- *Bacillus thuringiensis var. kurstaki*
- *Bacillus thuringiensis var. galleriae*
- NPV *of Helicoverpa armigera*
- NPV of Spodoptera litura
- Cymbopogan
- Neem based pesticides

(c) Successful implementation of biocontrol agents and biopesticides in Indian context

- *Helicoverpa* management on gram by applying NPV.
- Mango mealy bug and hoppers management and coffee pod borer management by *Beauveria bassiana.*
- DBM (diamond black moth) control through *Bacillus thuringiensis.*
- Sugarcane stem borer control by *Trichogramma.*
- Rot and wilt control of various cultivated crops by applying *Trichoderma.*

- White fly control cotton, chilli and okra by applying neem based products are effective

(d) Types of biopesticides

Biopesticides are classified into three different categories.

1. Microbial biopesticides
2. Biochemical biopesticides
3. Plant incorporated protectants

1. Microbial biopesticides: These biopesticides include bacteria, fungi, virus and algae, protozoa, etc. Microbial biopesticides are eco-friendly and less toxic as compared to inorganic pesticides. Aximum members of this group of pesticides are living microorganisms. Modes of action of these microbial biopesticides are of various types. Sometimes they secrete toxic substances when come in contact with the target organisms, sometimes they show hyperparasitism and several other modes of action. Microbial biopesticides are classified into three different classes :

- **Biofungicides:** Maximum members of fungi are plant or human pathogenic in nature. But there are several other members in this category who are non-pathogenic but show characteristic of entomopathogenic or hyperparasitism on different classes of pathogenic fungi. Therefore, such types of fungal biopesticides are effective against a number of insects and plant diseases. Entomopathogenic fungi are widely used for their eco-friendly, non-toxic nature which also show better lethal effect as compared to inorganic pesticides. The modes of action of entopathogenic fungi are very unique: attack hosts by entering through the cuticle, penetrating hemolymph and producing toxic chemicals which degrade the nutrients present in the haemocoel for insects self-immunity. Several biofungicidal products trigger plant defence systems, that can make plant more resistant against different stresses (biotic/abiotic). *Trichoderma* spp., *Beauveria bassiana, Metarhizium anisopliae* are used as biofungicides and are applied in nursery, ornamental crop, field crop, vegetable crops, etc. *Trichoderma* is applied for soil-borne, foliar diseases against different types of plant pathogens like *Pythium, Rhizoctonia, Fusarium, Sclerotium, Macrophomina, Phytophthors, Ceratobasidium,* etc. *Beauveria bassiana* and *Metarhizium anisopliae* are entomopathogenic fungi and control whitefly (including Q biotype) aphids, thrips, etc.

- **Bacterial biopesticides:** Bacterial biopesticides help to control several plant diseases and pests. Both sporogenus and non-sporogenus bacteria take part in controlling of insects and diseases. Endotoxins secreted by bacteria actively participate in controlling destructive insect/pests and diseases. The widely used bacterial strain is *Bacillus thuringiensis,* which shows lethal action on Lepidopteron insects. Approximately 40,000 species of *Bacillus thuringiensis* are used to control Lepidopteron pests. It produces Bt endotoxins by which they are attached with the digestive tract, stopping feeding and ultimately turn to mortality stage. Another two important bacteria are *Bacillus subtilis* and *Pseudomonas aureofaciens,* which have the potential to control damping off and soft rot disease causing pathogens. *Agrobacterium radiobacter* strain also has the potential to control *Agrobacterium tumefaciens* which causes crown gall disease. It is estimated that approximately more than hundred Bt based strains or biopesticide are effective against dipteran, lepidopteran and coleopteran larvae. One important entomopathogenic bacteria is *Serretia* sp., which is also considered to be a bacterial biopesticide.
- **Viral biopesticides:** Bacteriophages infect pathogenic bacterial cell and are biopesticide in nature. This helps to control rice, cotton and lepidopteron pests of vegetables. One of the widely applied biopesticide is NPV (nuclear polyhedrosis virus). Baculovirus strain *Anticarsia gemmatalis* nucleopolyhedro virus (AgMNPV), extracted from soybean, is an important biopesticide and is effective against velvet been caterpillar. Most of the caterpillars belong to the *Spodoptera* genus and are severe for agricultural crops. Two markedly available commercial viral biopesticides are Spodoptera NPV as SPOD-X ™ which contains *Spodoptera exigua* NPV to control destructive pests of vegetable crops and Spodopterin™ containing *Spodoptera littolaris* NPV which is applied to protect tomatoes, cotton and corn.

2. **Biochemical pesticides:** Different plant products specifically secondary metabolites like terpenes, flavanoids, alkaloids, phenols, suberins, carotenoids are also involved in developing plant immune system and show potential against both microbial pathogens and invertebrate pests. Combination of plant extracts and mineral oils can control targeted pathogen and manage other non-targeted microbes. Different plant extracts are also useful for stress (biotic and abiotic). As an example we can

say giant knotweed, *Reynoutria sachalinensis*, controls powdery mildew disease caused by *Oidium neolycopersiciin* and powdery mildew disease caused by *Leveillula taurica* in onion. *Artemisia judaica* actively controls Spodoptera littoralis, same as *Zingiber officinale* controls *D. melanogaster*.

3. **Plant incorporated protectants:** Plant incorporated protectants (PIPs) are those components which are easily insertable into plant genome. Cry protein of *Bacillus thuringiensis* inserted into different crops genome through genetic engineering. Transgenic brinjal (Bt brinjal) is such a Cry protein oriented product. Mode of action of Cry protein is different as compared to others like it is inserted into transmembrane pores which turn into osmotic cell lysis. Some of the common instances are brining Cry protein in maize hybrids (Cry 3Bb1), which is inserted to control *Dibrotica* spp. Another example is that cotton incorporated Cry1Ac protein is applied to control *Helicoverpa zea Boddie* (Lepidoptera), the expression of Cry 3A and Cry 3C protein was tested in potato to control *Leptinotarsa decemlineata* (Coleoptera).

(xii) **Application of pheromone trap for insect/pest management:** Pheromones are the substances present in living organisms and applied for sending chemical signalling to the opposite sex of the same species. These chemical agents are used in plant protection to control different types of insect/pest infestation to control substantial amount of crop losses. Compared to synthetic attractant, the success rate of pheromone is higher for trapping different classes of insects. Successful implementation of synthetic pheromone trap was tested on a number of horticultural crops. The application of pheromone trap was initiated during 1970s with the discovery of "hexalure" to control pink bollworm (*Pectinophora gossypiella*). Another such instance comes by the application of frontalin and trans-verbenol to control southern pine beetle.

(xiii) **Using of predators and parasitoids as biopesticides:** The predators usually kill and feed on the harmful insect/pests whereas parasitoids grow on or inside the host itself and live in their bodies and finally kill them. Examples of some predators are shown in Table 11.6. Parasitoids are found in Bihar hairy caterpillar (*Spilosoma oblique*) of jute plant and applied the same to control such insect/pests.

Table 11.6 Some important predators and their targeted pests

Sl. No.	Host	Predator	Target pest
1.	*Capsicum* sp.	*Amblyseius colemani*	*Scirtothrips dorsalis*
2.	*Phaseolus vulgaris, Solanum lycopersicon*	*Phytoseiulus* spp.	*Tetranychus urticae, Tetranychus evansi*
3.	*Solanum lycopersicon*	*Encarsia formosa*	*Bemisia tabaci*
4.	*Persea ameriacana, Capsicum annuum*	*Neoseiulus* spp.	*Tetranychus urticae, Oligonychus perseae*
5.	*Dendranthema grandiflora*	*Aphidius colemani*	*Aphis gossypii*

(xiv) **Using of bird perches to control insect pest:** Bird perches are the resting places for predatory birds to take rest and look for hawking of insect pests from different crops like cotton, peanuts and cowpeas. Maximum predatory birds prefer to look for prey in field crops if they get sufficient area for taking rest. To construct bird perches, use wooden poles or bamboo or tree branches. Maintaining such an arrangement gives good results. Example of such bird perches fulfill by bringing *Setaria* species (foxtail cultivars) in the crop field. These plants attract predatory birds. The birds feed seeds of this plant. At the time of cotton planting if it is maintained in every 9-10 th row of cotton, then it is very effective to control cotton bollworms and other insects.

11.2.8 IPM for Sustainable Agriculture with Magnification on ITK

Indigenous technical knowledge system (ITK) for the management of crops and our environment is in practice since long. Maximum ITKs continued through locally available components and human resources. ITK practices are eco-friendly and helpful for the sustainable farming system. Although with the increased demand regarding high yield converting traditional farming system to intensive farming system, where synthetic agrochemicals make a drastic change. Extensive use of hazardous, toxic, non-ecofriendly agrochemicals makes bad impact on destructive insect-pest-pathogen and their adoptive resistance. Based on this current scenario ITK is considered to be an important tool in organic farming and regain its position by consecutive use of the following practices.

11.2.8.1 Cultural Practices

(a) **Trap cropping:** Trap cropping is a traditional method of pest management. This technique is applied before the introduction of any disease from a certain region. In this management strategy extensive knowledge of pest management is required. In general, trap crop covers 10% of the total crop area. Example of trap crop like marigold/sorghum in tomato crop field in the ratio of 15:1, castor/sunflower in groundnut, pigeon pea in cotton, mustard in cabbage/cauliflower (1:30). Different crops like sorghum or pearl millet or maize act as a main barrier and attractant to the whitefly and aphids and restrict their entry.

(b) **Synchronous and timely showing or planting:** Timely sowing/ planting of disease-free seeds or seedlings or resistant varieties is an important practice to suppress the insect/pests. Early sowing of mustard (during first fortnight of October) of the short duration varieties under non-irrigated conditions escapes the attack of aphid and the leaf webber in the coastal districts of Odisha. Chickpea, sown between second fortnight of October and 1st fortnight of November, not only evades the outbreak of gram pod borer, Helicoverpa armigera, but also enhances the yield.

(c) **Healthy crop husbandry practices to avoid insect incidence:** In this category several steps are followed.

Plant material should be accurately examined for different insect infestations or mealy bugs before introducing into the crop. Sometimes it is observed that maggot infestation is controlled by Azolla cultivation. Stagnant water in rice fields may increase the BPH/WBPH infestation rate, so time to time, draining of stagnant water is required. Try to avoid heavy insecticides spraying at the time of high parasitoid activity. Excessive nitrogenous fertilizers can control phloem feeding pests like aphids. Spiders constitute an important portion of predatory arthropods in several ecosystems (Chaterjee et al., 2009). It is found that 90% of the harmful pests and insects are controlled by spiders in rice plots.

(d) **Enhancing/conserving natural enemies:** Ideal enhancement of habitats make sure the survival and improvement of natural pests by providing barrier for their augmentation. Planting various plant species or a range of sequentially flowering species inside or closest to the fields provides the natural pests a secondary host source. This makes the alternative attractant to the natural enemies. It is suggested that 20% area of the peripheral cropland is planted with *Glyricidia sepium, Cassia auriculata,* drumstick, karanj, neem, jatropha, etc. These vegetations can preserve both pollinators and recommended arthropods. Several trap crops also help in controlling insect/pest populations by enhancing the natural enemies within the field. Sorghum is used as a trap crop to control cotton bollworm, *Helicoverpa armigera,* that also increase rates of parasitism by *Trichogramma chilonis* (Virk et al., 2004).

11.2.8.2 Physical or Mechanical Control

This process involves the mechanical removal of pests. Such as hand-picking the larvae and grubs, removing eggs from the tips of the leaves by pinching off the terminal portion, by creating flames with bonfires and several different other techniques. Certain local indigenous techniques found to be most effective and used for suppression of destructive insect pests. Tribal women in northern Odisha pick tenden terminal leaves during early vegetative stage of chickpea crop which is a source of income for them. It has been observed that irrespective of showing dates and varieties, *Helicoverpa* preferred the 2nd and 3rd leaf for oviposition in their lifecycle on chickpea. Therefore, the practice of nipping tender leaves by tribal women decreases the *Helicoverpa* population. Schematic representation of IPM principles and allied methods are shown in Fig. 11.2.

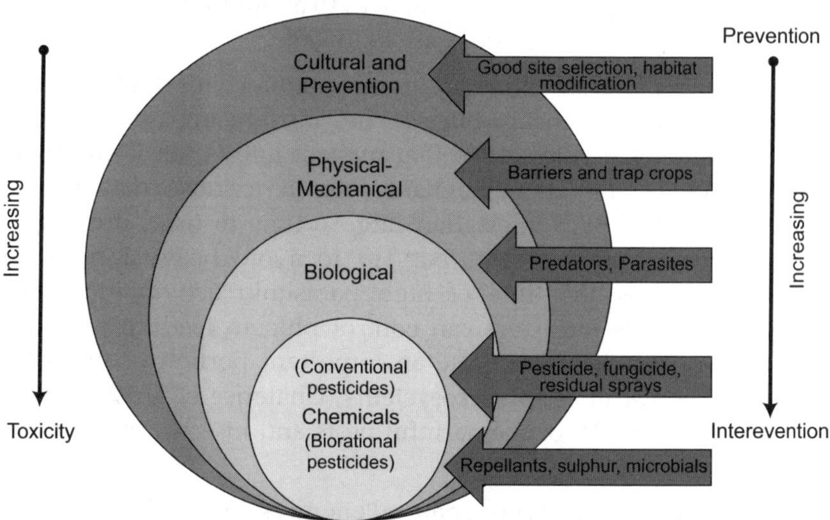

Fig. 11.2 Schematic representation of IPM principles and allied methods.

11.2.9 Other Control Methods

Several instances can be mentioned under this category. First, Mucuna prpuriens (Farm: Papilionaceae) fruits are retained in the rodents' hole. When it comes in the contact of this hairy fruit , rodents suffer from irritating hair and leaf spots. Secondly, mixture of 90% crushed seasame or groundnut or niger with addition of 5% sugar and 5% powdered formed fused electric bulb or tube crushed and prepare as food for the rodents food. With this mixture rodents die within few weeks. Stem portion of Jatropa plant (10-12 inches long) cut into small pieces is good for rat control. Rat infestation on coconut tree is controlled by using palm leaf attached it dorsiventral surface with the stem portion of coconut tree trunk.

11.3 TRAPPING INSECTS WITH INDIGENOUS TECHNIQUES

Trapping of destructive insects pests by using different indigenous techniques is another site of IPM. Some of the techniques are described below.

(a) **Trapping of rhinoceros beetles:** Powdered form of castor cake (250 g) is mixed with 1/3 amount of water and poured in a clay pot. This pot is kept under the soil. The attractive smell brings the beetle near the pot which gradually falls into the water. Placing of such types of pots (3 or 5 in number) per hectare coconut plot can control the rhinoceros beetles. Sometimes, pineapple traps are used to catch the rhinoceros beetles. In this system few slices of pineapple are kept in the cylindrical pot in which an exit hole is made to allow the rainwater to flash out. This type of trap is hung near the crown of a coconut tree.

(b) **Trapping of fruitflies:** Fruitfly incidence is mainly found in Mangiferaceae and Cucurbitaceous members. Various indigenous techniques are used to control the fruitfly infestation. In a coconut shell 20 g of crushed *Ocimum sanctum* leaves are taken and mixed with 100 mL water. In this solution 0.5 g of citric acid and 0.5 g of carbofuran 3G is taken. This trap is placed in 4-5 in number per tree. Fruitflies consume this extract and die. Another example is that in a 2 litre disposable water bottle 1 cup amount of vinegar, 2 cups of water and one table spoon honey is taken. Two holes are made on the bottle 5 cm apart as an entry point of the insects. Place the traps on the tree trunk (approx 5 ft height). Flies will come with the attractant solution and trapped.

(c) **Trapping of gundhi bugs in rice:** By using dead crabs, fogs or pieces of jackfruit and attaching it with bamboo sticks is placed in rice fields before milky stage come. This acts as an attractant to gundhi bugs. Burning of rubber tyre catches gundhi bugs easily.

(d) **Trapping of red palm weevil (RPW):** The midrib portion of the coconut leaf is cut into small pieces and transferred into an earthern pot with 1 litre water, 100 gm jiggery, 10 gm tobacco powder, 2.5 kg sugarcane molasses, 5 mL acetic acid, 5 gm yeast powder. Another pot is placed at its bottom over it. This type of arrangement will set in different corners of coconut orchards. Time to time jiggery, tobacco and water should be changed if it dries.

(e) **Trapping of blister beetles:** It is found that 1.5 litres of water added with 10 gm of detergents is very useful, when kept in blue containers for trapping blister beetles.

(f) **Trapping/sucking insects/pests:** In this trapping technique bright yellow sticky traps are used for controlling aphids, thrips and whiteflies. For monitoring whitefly, thrips, etc. Ten yellow sticky traps are set apart per hectare on cropping land. Disposable tins and glasses are printed in yellow by preparing yellow coat through white grease, vaseline and castor yellow. Place the sticky trap approximately 25 cm apart from the main plant trunk. Ensure that the leaves do not stick the board and not face direct sunlight. Hang the traps 50-75 cm above the plants. Same as yellow water pan traps placed for controlling and monitoring of aphid populations.

(g) **Trapping of slugs in kitchen garden:** In an igloo style rinds of grapes, set by leaving small amount of pulp on the upper side. The slugs generally hide underneath of the grapefruit and die.

11.4 USE OF BOTANICALS AGAINST INSECT PESTS

Different plant parts are used as traditional medicine against different insect pest infestations. Botanicals are easily available and cost effective. In tribal and villege agri faming they prefer botanicals in place of inorganic pesticides. For example, traditional farmers in Odisha believe that to control of leaf folder, gallfly and caseworm, the best control measure is to apply karra or karada leaves (*Cleistanthus collinus* / Euphorbiaceae) in rice fields. The applied doses are about 50-60 twigs of 4-5 feet long / acre or in another way 10 kg leaves are applied for 1000 square metre. Farmers of western Odisha apply one technique to control yellow stem borer, leaf folder, and case worm. They make a mixture by adding jaggery (50 gm), cow urine (5 lt), fresh cow dung (1 kg), neem leaves (1 kg), *Calotropis* leaves (1 kg), karanj leaves (1 kg) and keep 1-2 weeks for fermentation. After proper fermentation the surface liquid is collected and applied at the rate of 15-20 mL per litre of water. There are several other botanicals which are used against different insect pest infestations like 2% tulsi extract, seed and lemon grass (10-30%), dhatura extract (1%), neem oil (1%), *Coriandrum sativum* (against spider mites) and *Vitex negundo* (against diamond black moth, hairy caterpillar, leaf folder, rice stem borer, semilooper) solution, etc.

11.5 USE OF DAIRY PRODUCTS TO CONTROL PEST

Cow milk can be used as an effective blocker against different diseases because of the presence of casein protein. It can be applied with 10% aqueous suspension against powdery mildew. Spraying of milk generates systemic acquired resistance against chilli leaf curl virus. It has also been observed that curd water is used for the control of white fly, jassids, aphids, etc. Spraying of well fermented buttermilk (2 days) at the rate of (1:9) ratio of fermented buttermilk and water is very effective to control whiteflies in okra plant (Karthikeyan et al., 2006).

11.6 USE OF ASH FOR PEST CONTROL

Different types of ashes are rich source of phosphorus and also act as a physical poison which causes abrasion on insect's body cuticle and insect dies due to desiccation. Spraying of this ash particles near the crop field protect the farm from different kinds of pests. This type of ashes are effective against cutworms, caterpillars, grasshoppers, etc. Several instances like application of ash 50 kg/ha, kerosene 5 %, and spinosad 45 SC mixture used in the ratio of 4:8:1 in brinjal (Patnaik and Dash, 2017). Similarly, rice seedling beds treated with fly ash control destructive insect pests.

11.7 USE OF KEROSENE FOR PEST CONTROL

Using of kerosene-soap water emulsion in crop field and application of this mixture as an insecticide is an old concept (Jex-Blake, 1950). This mixture/solution is effective against scale insects like bugs, mites, aphids, leaf minors, etc. (Van der Werf, 1985). At higher concentration it shows phytotoxicity so use of foliar spray should be restricted between 1-2%. Kerosene is easily available and requires in very little amount is considered alternative of chemical pesticides.

11.8 SUCCESSFUL EXPEDITION OF IDM FRAMEWORK

11.8.1 Construction of Plant Protection Policy

(a) **Ideal implementation of plant disease and pest management policy:** To reduce the crop loss both in qualitatively and qualitatively, Ministry of Agriculture is implementing several policies like disease pest forecasting, disease surveillance, information sharing and decision support system, etc. All the policies are mainly beneficial while some policies can be harmful or hazardous in some cases like regular application of pesticides, free distribution of chemicals to farmers, etc.

(b) **Systematization of eco-friendly pesticide management policy:** Policies and regulations are implemented to check the pesticide toxicity level, its registration and labelling, guidelines for application, skilled personnel, etc. Banning or limited application of hazardous chemicals should be implemented to lessen the harmful effects on farmers as well as environment. Some countries have removed import tax and employ pesticide subsidy to promote less hazardous agricultural pesticides which can be added to IDM practices.

(c) **Establishment of plant quarantine:** Plant quarantine is one of the major steps to monitor the introduction of new pathogen or pest to an affected area to healthy area. Different quarantine stations should be established in the point of entry to check the planting materials or other agricultural products. Citrus tristeza virus, once attacked citrus plantation in many countries. Its vector is added in quarantine regulation and control (Moreno et al., 2008). Pest risk analysis (PRA) and pest disease survey are the two important steps to prepare and update quarantine pest list.

(d) **Implementation of seed certification and vegetative propagation policy:** Using quality and healthy seeds and planting materials is one of the major approaches in IDM. If the planting materials are not checked properly before sowing or grafting, they can transmit many viral, bacterial and fungal diseases. So the government should implement and regulate polices related to cost, quality and availability of planting materials. Certified seed production fields should be monitored and inspected regularly to check their quality. Policies have been implemented for not to multiply hybrid and GMO seeds as their viability may deteriorate.

(I) **Proper planning of private sector investment policy:** Private sectors are involved in IDM practices for providing inputs and resources to farmers like quality planting materials, bio-pesticides, growth inducing chemicals, biocontrol agents, pheromone and yellow sticky traps, etc. The involvement of private sectors may be critical if they don't supply the above-mentioned products qualitatively. IDM and IPM approaches can be different according to the location and one major cause to limit the involvement of private sectors is to reach the remote villages. Government should take steps to implement incentives and reduce the tax which may help the private sectors to work properly in these areas (SP-IPM, 2008).

(II) **Development of proper trade and export policy:** IDM policy at national and world level depends on each other through trade and export. Diseases transferred from one country to another by travellers or through other means are very problematic and the countries should cooperate and update any recent spread of disease, pest or its any virulent strain before it causes pandemic. Global wheat rust monitoring system at FAO and late blight disease, European network, Euroblight are two examples of such system.

(III) **Implementation of Act for public health, food safety and environment:** The implementation of this Act is most important to preserve the safety of humans and environment by maintaining the pesticide residues and other toxic contaminants in food and waterbodies. Regulatory system for limited use of GMOs and pesticide can reduce the risk for environment. In developing countries, it may be difficult to form a regulatory body as it requires more human involvement, finance and lack of co-ordination in information sharing, and proper infrastructure.

(IV) **Planning of land use policy and its implementation:** The land ownership and its size are limiting factors for the farmers to follow any new technologies with long-term return or IDM Management of alternate or collateral hosts, natural enemies, monocropping, less varieties grown in agro-ecological region, crop genetic diversity, etc., may directly or indirectly affect the effectiveness of IDM.

(V) **Establishment of education and awareness policy:** Awareness should be created and farmers should gain knowledge on IDM, its different approaches, benefits, etc., by involving NGOs and other organizations. Framers can be more aware about IDM if they follow the method of "learning by doing". The government should take steps forward to aware the farmers through different policies, regulations and funding (van de Fliert et al., 2000)

(VI) **Ramification of agricultural extension policy:** IDM approach should reach through "diffusion of innovation concept" from top scientists to farmers by extension workers, agriculture and horticulture officers. The extension officers should visit and train the farmers. Trained farmers explain to other community members and the knowledge moves through.

11.9 CONCLUSION

Day by day growing interest in organic farming and its different tools is a positive approach for us. Research on the present scenario revealed that farmers gained more advantage from organic agriculture in the less rainfall, low soil fertility level area. Integrated disease management is an important tool of organic farming. By adopting such tools farmers gain new techniques and skills to continue their farming within farming budget. IDM practices not only protect our crops but also give more production. From this point of view we can conclude the following points.

1. Large-scale conversion from inorganic to organic agriculture enhances crop production, future security and sustainable environment-friendly approaches.
2. Organic manure is the best alternative renewable source among all types of market based inorganic products enriched with high nutrient.
3. For proper utilization of IDM strategies, knowledge should be distributed among economically poor and illiterate farmers who are not aware about this techniques.
4. Skill transfer is needed in terms of organic products certification and uses.
5. Knowledge about destructive insect-pest-pathogens is highly needed. Crop calendar and their alteration in terms of management is also an important aspect. Epidemiological knowledge and behavioural characteristics of insect-pest-pathogen is necessary for better security of crops.
6. Government help and subsidiary is required to encourage organic growers.
7. Meeting, leaflet, workshop or digitalized app based skill transfer on IDM strategies makes the organic movement more fruitful.
8. Distribution of disease-resistant and stress-resistant seeds from government is a good approach to encourage farmers in developing countries.
9. Use of biopesticides on the cropland and recommended doses should be strictly followed.
10. For more successful IDM strategies, selection of farming land, proper irrigation, time to time weeding and cleaning is a good farming practice.

11.10 QUESTIONS

Short Questions

1. What is crop rotation? How it is important in organic farming?
2. What is biopesticide? Give examples on botanical, fungal, bacterial and viral biopesticides.

3. What is pheromone trap? Describe its significance.
4. What is Cry protein? Mention its source of extraction and its mechanism.
5. How disease-resistant variety helps in organic farming.
6. What is bird perches? How is it maintained in crop fields?
7. What is biocontrol agent? How does it work? Give example.
8. Define PGPR. Give examples.
9. Describe plant incorporated protectants (PIPs). How does they work?
10. What is trap crop? How does it help IPM?

Long Questions

1. What is IDM? Mention its principle and significance in Indian context.
2. Describe disease management strategies applied in organic farming.
3. What is biopesticide? Describe different groups of biopesticides and their advantages.
4. Describe ITK based integrated pest management for sustainable agriculture.
5. Describe successful implementation of IDM in Indian context.

REFERENCES

Chatterjee S., Isaia M. and Venturino E., 2009. Spiders as biological controllers in the agro ecosystem. *Journal of Theoritical Biology,* 258: 352-362.

Jex-Blake A.J. 1950. Gardening in East Africa, Longmans Green and Co. Ltd, London.

Karthikeyan C., Veeraragavathatham K.D. and Ayisha F. S. 2006. Cow based indigenous technologies in dry farming. *Indian Journal of Traditional Knowledge,* 5(1): 47-50.

Litterick, Audrey M, Watson, Christine A and Atkinson David., 2002 Crop protection in organic agriculture- a simple matter? Published by Powell J et al., Proceedings of the UK organic research 2002 Conference, pp. 203-206. Organic Centre Wales, Institute of Rural studies, University of Wales Aberystwyth.

Moreno P., Ambros S., Albiachi-Marti M.R., Guerri J., Pena L., 2008. Citrus tristeza virus: a pathogen that changed the course of the citrus industry. *Molecular Plant Pathology,* 9: 251-268.

Patnaik H.P. and Dash S. 2017. ITK based IPM for sustainable agriculture. Published souvenir of International conference on organic farming and sustainable agriculture (OFSA), pp. 26-34.

Van de Fliert E., AsmunatiR.,.Tantowijoyo W. 2000. Participatory approaches and scaling-up. In: Stür W.W., Horne P.M, Hacker J.B., Kerridge P.C. (eds). Working with farmers: the key to adoption of forage technologies. ACIAR Proceedings. No. 95, International Workshop, Cayagan de Oro City, Mindanao 1999: 83-90.

Vander Welf E. 1985. Pest management in ecological agriculture. AME foundation, Groenekan/Holland. Vijayalakshmi K., Subhashini, B. and Shivani, V.K. (1999). *Plant in Pest Control: Garlic and Onion*, Centre for Indian Knowledge and System, Chennai, pp. 1-20.

Virk J.S., Brar K.S. and Sohi A.S. 2004. Role of trap crops in increasing parasitation efficiency of *Trichogramma chilonis* Ishii in cotton, *J. Biol. Control*, 18: 61-64.

12 CHAPTER

Weed Management

12.1 INTRODUCTION

Organic farming is the best substitute of conventional method of farming where more chemical is used and considered less eco-friendly. Though the yield gap between both systems of farming is very less, still organic farming is in high demand due to its eco-friendly nature. But one of the major constraints in organic agriculture is weeds. As the application of chemical herbicides is banned or restricted in organic system, proper management and care is required to control weeds in the field. This chapter gives a brief description on management of weeds in organic system in an eco-friendly way.

12.2 WEED: WHAT IS IT?

These are the unwanted plants or grasses other than the main crop in a field. These are one of the most omnipresent groups of crops resulting heavy crop loss. These compete with main crop for water, mineral, nutrition, etc. They also deteriorate the quality of food by decreasing the quality. Researchers have reported that 30% of crop yield loss is due to weed problem (Barberi 2002). Therefore, ensuring food security for the increasing population can only be possible by controlling weed.

12.3 CRITICAL PERIOD OF WEED CONTROL

It is the time period in a crop life cycle where the field should be weed free. If not, then maximum yield loss occurs. The weeds which grow after this critical period do not affect much on the crop yield. But controlling weeds even after critical period results in proper harvest of crop, good quality of produce, keep the field disease- and insect-free, and less weed infestation problem in subsequent crops (Gomiero 2011). Horticultural crops are most vulnerable to weed infestation and require proper care from planting till critical period is over. Critical period for weed control is listed in Table 12.1.

Table 12.1 List of horticultural crops with their critical periods of weed control

S. No.	Crop	Stage of crop	Critical weed control period
1.	Tomato	Newly planted	36 days after transplanting
		Seed bearer	9 weeks after seed bearing
2.	Potato		4 weeks after planting
3.	Onion		From planting to harvesting
4.	Lettuce		3 weeks after planting
5.	Cucumber		4 weeks after seeding
6.	Carrot	Newly planted	3-6 weeks after emergence
7.	Beet		2-4 weeks after emergence

Weed management in organic system restricts the use of chemicals to very few extents. Some of the eco-friendly practices are illustrated below.

12.4 CULTURAL METHOD

12.4.1 Crop Rotation

Crop rotation refers to alternation of the crop with another crop having different growth characteristics. Normally weeds choose the crop according to the growth characteristics as well as the same nutritional requirement. Therefore, in crop rotation the main crop and the subsequent crops would be chosen in such a way that both have different growth characteristics. In case of monoculture the same crop is grown year after year for a long period of time in the same field. So more weed infestation of same species is observed in the field but in case of crop rotation, the cultural practices like tillage, weeding, mulching are different from crop to crop. Hence, disrupts the life cycle of the weed and lowers its infestation.

Choosing the suitable crops is one of the important criteria for the grower as it decides the weed infestation in the main crop as well as in the following crops. It was observed that closed planted crops with row crops, heavy feeders with light feeders, legumes with grasses are some of the suitable crop rotations to lessen the weed infestation. The land should we kept fallow for a period of time to reduce the population of perennial weeds. This, is also known as fallow period.

12.4.2 Cover Crops

Cover crops such as rye, buckwheat, winter wheat, red clover, forage crops grow rapidly and cover the farm area with their dense population. Due to the crowded covering, weeds do not grow in the farmland. Their residue also helps in cooling and shading of the soil and keep the land free from weeds. It was observed that when the residue of cover crop is mixed with soil, several allelo chemicals are secreted resulting in inhibition of weed germination.

12.4.3 Intercropping

Intercropping also plays an important role in controlling the weed population. One of the major focus areas in intercropping is choosing the main crop and intercrop. It would be chosen in a way that both should not compete for water nutrition or any minerals. The intercrop should be grown in between the rows of the main crop.

12.4.4 Field Scouting

Field scouting is one of the crucial steps to collect data and get acquainted with different types of weeds, their stages, crop stages, and critical period of weed control. If monitoring is done at proper time intervals, then severe weeds can be removed before the critical period.

12.4.5 Mulching

Mulching or covering of the main field is utmost important as it blocks the sunlight and secrets some allelo chemicals. This reduces the germination of weed seeds. Mulching is normally of two types: one is live mulch and another is organic mulch.

Live mulching: This type of mulching includes plant species which normally grow near the soil zone such as *Portulaca olerasia*. The principal purpose of growing this type of plants is to conserve water, minerals and nutrients and to stop pest as well as weed infestation. These crops should be grown just before planting of the main crop or after harvesting. It should not be grown with the main crop simultaneously as it may compete for water minerals and nutrients. It was identified that broccoli can act as mulching crop without competing with the main crop.

Organic mulching: Organic mulch mostly includes woods, barks, crop residues, fallen leaves, straw and any other composting material. Mulching by these materials helps to keep the farmland weed free because proper light cannot penetrate in the soil. It is advised to grow all the composting material in own farmland which will be cost effective. organic mulches available in market have exorbitant price. Organic mulching can be done by using newspapers and straw. In this method, two layers of newspaper and straw cover the soil which keeps the soil cool, light free and hence, weed free. Black polythene mulch which is normally available in market is also an alternative of composting material. This black polythene mulch doesn't penetrate any photosynthetic radiation but lets infrared light to go in which keeps the soil warm and kills all type of weeds and weed seeds.

12.4.6 Selection of Crop Variety

Careful selection of variety not only keeps the field weed free but also decreases any other pest problem to a great extent. Varieties should be selected which have rapid growth characteristics as compared to the weeds.

12.4.7 Tillage

The tillage operation helps to control the weeds by changing the soil seed bank and also modifying the depth of the seeds. It was observed that tilled fields are less infested by weeds as compared to untilled fields.

12.4.8 Field Sanitation

Field sanitation is one of the important factors to control the weeds. For this purpose, we have to use weed-free seeds. Proper monitoring is required to eradicate the weeds and its seeds which will prevent any future weed infestation. The weeds should be removed from the land as well as outside of the land. Movement of animals should not be allowed because they act as carriers of weed seeds. Hand weedings is also required. The machinery or implements used in the land should we cleaned properly before using in a new field.

12.4.9 Fertilization

The growth of weeds also depends on the type of fertilizer, its application, and the dose. For example, nitrogen fertilizer is more preferred by the weed species than any other crop. So, it's better to choose the legume residues or green manure crops as the source of nitrogen rather than any chemical fertilizer. Pre-planted broadcasting of fertilizers should be avoided as it stimulates the weed growth rather than crop growth. Fertilizer should be applied near the plant roots or near the row. Organic fertilizers are recommended as they decrease the weed population.

12.5 MECHANICAL CONTROL

Weed management by using different implements such as hoes, plough, harrow, etc., is one of the effective methods of weed control. Though it is labour intensive or time consuming, still it gives the best results. Soil seed bank is the main factor in weed infestation. Therefore, the main aim to modify the seed bank in a way that weeds or its seeds don't grow successfully. When a land is ploughed properly, it cuts the weeds from its root portion or bury the seeds to a depth that it won't germinate. Blind cultivation is one of the effective methods which controls the weeds, its seeds and just germinated weeds without hampering the main crop.

12.6 BIOLOGICAL CONTROL

Biological control refers to managing the weeds by using several biocontrol agents or living organisms which are also known as predators, parasites etc. It includes insects, microorganisms, birds or animals or any other crop species. Mycoherbicide is one formulation containing spores of a fungi to control weeds and applied as a spray. Likewise, a caterpillar (*Cactoblastis* sp.) is used to control the prickly pear. Still more research is going on to discover many species of predators and parasites to control the weeds (Kremen 2012).

12.7 ALLELOPATHY

Allelopathy can be defined as direct or indirect chemical effect of one crop upon another. The chemicals secreted which affect other is known as allelo chemicals. This hampers the other crop growth and germination. A wide range of plants such as rye, buckwheat, carrot, radish, oat, alfalfa, etc., show allelopathic effect. It was observed that growing of buckwheat before six weeks of planting not only restricts the weed growth but also stores phosphorus in soil

12.8 HEAT TREATMENT

Heat treatment, using different types of flamers, ionized radiation, is one of the potential methods to control a wide range of weeds. Flamers are normally used before the crop has emerged or after the crop has emerged. In this method the cell sap expands and it bursts the cell. But sometimes, this method also hampers the main crop. Hence, this method is not advised to farmers. Soil solarization is one of the effective methods to control the weeds. In this method not only the weeds but also the harmful microorganisms, insects and nematodes are killed. The soil is exposed to solar radiation for a longer time period. Then it is covered by black polythene to keep the soil heated. The increased temperature helps to kill the pest.

12.9 CONCLUSION

Weed management through organic ways is one of the best effective methods to sustainable agriculture. Though chemical is restricted to some extent in organic weed management, but cultural, thermal and biological management help to control the weed successfully. More awareness programme is still required to deliver this methods to all rural farmers.

12.10 QUESTIONS

Long Questions

1. What do you mean by weed? List out the major weeds affecting cereal crops with their organic ways to control.
2. Explain the cultural method of weed control.
3. Describe the weed control through heat treatment.
4. How weeds can be managed by using biological agents.
5. How organic weed management is more beneficial than chemical method.

Short Questions

Write Short Notes on:

1. Allelopathy
2. Soil solarization
3. Tillage operation
4. Crop rotation to control weed
5. Mycoherbicide

REFERENCES

Barberi, P. 2002. Weed management in organic agriculture: are we addressing the right issues? *Weed Research*, 42(3): 177-193.

Gomiero, T., Pimentel, D. and Paoletti, G. 2011. Environmental impact of different agricultural management practices: conventional vs. organic agriculture. *Critical Reviews in Plant Sciences*, 30(1-2): 95-124.

Kremen, C. and Miles, A. 2012. Ecosystem services in biologically diversified versus conventional farming systems: benefits, externalities, and trade-offs. *Ecology and Society*, 17(4): 40.

13 CHAPTER

Operational Structure of NPOP

13.1 INTRODUCTION

In the era of modernization, agriculture is also shifting towards new technologies. Recent trends are preferred for sustainable development in an eco-friendly way. Organic farming is one of the important systems to reach sustainability in agriculture. The market as well as the consumer demand for organic products in increasing due to the low or no chemical residues (Singh, 2009). To maintain the trust of the consumers and quality of the product, several regulatory bodies are structured to maintain the rules and regulations. Without the regulatory bodies, the marketability of organic products may not be maintained by the consumers as organic farming system preserves the quality of the final produce by checking it from land preparation to selling. These regulatory bodies form rules, regulations, programmes to improve the standard of organic produce.

13.2 NATIONAL PROGRAMME ON ORGANIC PRODUCTS (NPOP)

Ministry of Commerce and industry, Government of India started a programme called NPOP (National Programme on Organic Products) in the year 2000 and implemented in 2001. It comes under Foreign Trade and Development Act. NPOP is implemented under AGMARK by Ministry of Agriculture, Government of India. The rules and regulations of NPOP matches with the standards of IFOAM and Codex. NPOP has analogues agreement with European Union and Switzerland. USDA has also accepted the conformity system of NPOP (Charyulu and Biswas 2011).

13.3 AIMS OF NPOP

1. To provide the means of evaluation of certification programme for organic agriculture and products (including wild harvest, aquaculture, livestock products) as per the approved criteria.
2. To accredit certification programmes of certification bodies seeking accreditation under this programme.

3. To facilitate certification of organic products in conformity with the NSOP.
4. To facilitate certification of organic products in conformity with the importing countries' organic standards as per equivalence agreement between the two countries or as per importing country's requirements.
5. To encourage the development of organic farming and organic processing.
6. To advise the National Steering Committee on relevant issues in National Standards and Accreditation.
7. National Accreditation Policy Programme is administered by the National Accreditation Body, which defines the overall policy objectives for the accreditation programmes and operations.

13.4 SCOPE OF NPOP

The NPOP shall, inter alia, include the following:

(a) Policies for development and certification of organic products as notified by the Department of Commerce from time to time
(b) National standards for organic products and processes.
(c) Accreditation of certification programmes to be operated by certification bodies.
(d) Certification of organic products.

13.5 WORKING COMMITTEES FOR NPOP

A committee is formed to monitor all the procedures from the time of land preparation to the selling of the product. Proper inspection and certification are required by govt. appointed agencies to monitor this. This committee is accredited by national accreditation body. This committee includes some members of Ministry of Agriculture, APEDA, Coffee Board, Spice Board, Tea Board, Export Inspection Council of India (EIC) and inspection agencies (EIAS) (Fig. 13.1).

1. **Department of Commerce:** Department of Commerce is the apex body for national programme for organic production. NPOP Works under the guidance and direction of this organization.
2. **National steering committee:** This is the apex policy formulation body set up by Department of Commerce, government of India. This committee is headed by Commerce Secretary. He can nominate any other member to head the meeting. Implementation and administration of NPOP is the major responsibility of this committee. The members of this committee are chosen from

Department of Commerce, Ministry of Agriculture, Ministry of Textile, Ministry of Animal Husbandry, Dairy and Fishery, Ministry of Food Processing Technology, Ministry of Science and Technology, Ministry of Environment and Forest, APEDA, MPEDA, FSSAI, Spice Board, Coffee Board, Tea Board, etc. This committee should meet time to time at least once per year to amend new laws/rules or to monitor the procedure of NPOP. For this meeting the strength should be minimum 30%. This meeting can include other members or sub-committee bodies for the smooth functioning of the NPOP regulations.

3. **National accreditation body:** National accreditation body works under APEDA. It includes the members of Department of Commerce, Ministry of Agriculture, FSSAI, MPEDA, APEDA, Coffee Board, Spice Board, Tea Board, etc. The Additional Secretary of plantation heads the meeting. This body should meet according to the requirement. This organization draws

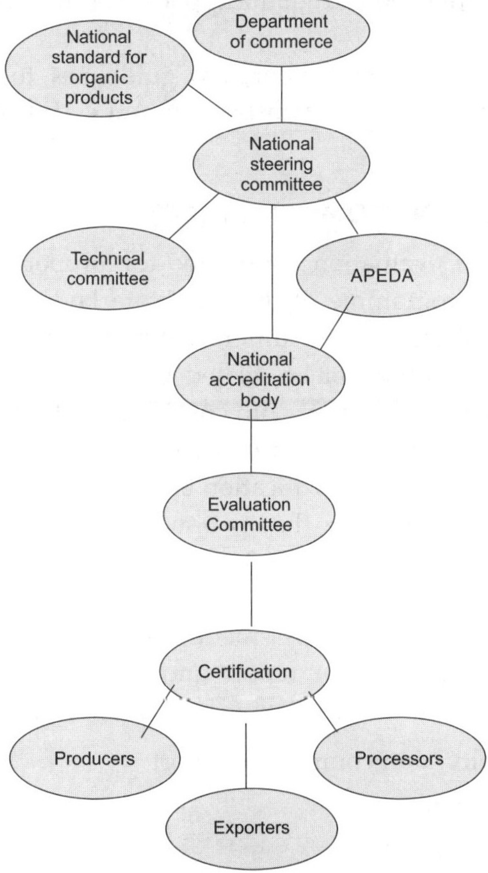

Fig. 13.1 Operational structure of NPOP.

of the acquisition of certificate bodies, the procedure and evaluation system of certificate bodies and constitute the evaluation committee.

13.6 AGRICULTURAL AND PROCESSED FOOD PRODUCTS EXPORT DEVELOPMENT AUTHORITY (APEDA)

APEDA is well recognized as the secretariat for the implementation of the NPOP. The responsibilities of APEDA include the following (APEDA 2006):

- Implementation of the decisions of the NSC, NAB and the committees constituted under the NPOP.
- Organize meetings under NPOP.
- Evaluation of the certification bodies.
- Investigation of grievances acknowledged from the importing countries.
- Screen applications from the applicant bodies and arrange their evaluations.
- Issue required implementation guidelines to the accredited certification bodies for inspection and certification from time to time

13.7 EVALUATION COMMITTEE (EC)

NAB sets up an evaluation committee for proper implementation of certification programme of the certificate bodies. It includes the members from the experts of different agriculture and allied fields. These members should not be related with any of the certification bodies and sign a contract with APEDA to keep their identity confidential. Proper training is provided to the experts for monitoring and inspection. The same certification body shall not be evaluated by the same committee for more than two successive years. An evaluation committee comprises minimum 3 experts from different fields and submit the documents to APEDA. The following documents are submitted to APEDA for assessment. APEDA reviews the reports and submits its assessment report and present it, along with its recommendations, to the NAB for accreditation decision.

(i) Conformity/non-compliance report

(ii) Observations

(iii) Recommendations

(iv) Supporting documents

13.8 TECHNICAL COMMITTEE

The NSC constitutes various technical committees including experts from relevant organizations to frame various technical standards, suggest amendments, review the standards from time to time and recommend the NSC on relevant issues relating to organic sector.

13.9 CERTIFICATION BODIES

These are agencies accredited by the National Accreditation Body under NPOP for certifying organic products. The accredited certification bodies certify organic products as per the choice of accreditation approved by the NAB (Ummyiah et al., 2017).

13.10 NATIONAL STANDARDS FOR ORGANIC PRODUCTION (NSOP)

NSOP refers to the production, processing, handling and labelling standards of the products coming in certain categories given below:

(i) Crop production

(ii) Livestock, poultry and products

(iii) Beekeeping /apiculture

(iv) Aquaculture production

(v) Food processing & handling

(vi) Any other category of products that the NAB may include from time to time

13.11 CONCLUSION

In organic production system, certification is one of the major steps and it is the mandatory stage for marketing of the products. Without proper certification, the whole produce will not be sold as organic. Therefore, all the procedures should be known by the farmers. Proper measures should be taken by the government to deliver the information about certification process, rules and regulations of certification so that both growers and consumers are benefitted.

13.12 QUESTIONS

Long Questions

1. What is NPOP? How does it work?
2. Explain the procedure of certification?
3. How an organic product can be marketed with proper labelling? Briefly explain.

4. Is NPOP important for organic production? Justify.
5. What is AGMARK and other logos used for organic products?

Short Questions

Write a Short Note on:

1. Certification body
2. Spices board
3. NPOP
4. APEDA
5. Evaluation committee
6. Trade mark
7. AGMARK

REFERENCES

APEDA. 2006. Accredited inspection and certification bodies. In National Program for Organic Agriculture (NPOP) - standards.

Charyulu, D.K. and Biswas, Subho. 2011. Organic input production and marketing in India-Efficiency, Issues and Policies. CMA publication no. 239, Allied Publishers. IIM, Ahmedabad.

Singh S. (2009) Organic produce supply chains in India: Organization and Governance, CMA publication no. 222, Allied Publishers. IIM, Ahmedabad.

Ummyiah, H.M., Afroza, N.J.B. Mushtaq, F. et al. 2017. Certification of organic products. *International Journal of Current Microbiology and Applied Science*, 6(11): 2300-2308.

Unit 5

Quality Maintenance, Certification and Marketing

14 CHAPTER Quality Considerations, Processing, Labelling, Economic Considerations, Viability and Accreditation Processors

14.1 INTRODUCTION

Organic farming is gaining more attention in recent period throughout the world. Due to its eco-friendly and no or low residue effect, the demand is increasing. Customer satisfaction and faith on this product is one of the major factors for its success. But many companies are out of the track and selling products as organic without any certification. Thus, government has made regulations to accredit some agencies who will certify the products with time to time inspection and monitoring. A manual is prepared by the competent authority where all the standards, procedures and permitted chemicals with doses are listed.

14.2 QUALITY CONSIDERATION

Quality of organic products is one of the key factors for its high market value. Before going to open market, the produce is sent to the certified labs for testing its quality and any chemical residue. Hence, proper care should be taken to preserve its quality. Most of the time quality decreases at the time of processing to marketing due to more post-harvest contamination, adding of food preservatives, colouring agents, etc. The requires standards should be maintained during the processing which is described below.

14.3 PROCESSING

The quality of products should not be hampered at any stage of processing. Proper techniques should be followed with utmost care. Refining of products, use of additives and preservative chemicals should me minimized. Ionizing radiations should not be used to control pests and pathogens or for food preservation purpose. The specified standards are described below which should be maintained according to the stage of processing:

14.3.1 Pest Management

- Preventive methods should be followed. Any chemical treatment or radiation treatment are prohibited.

- If preventive method does not work, mechanical or biological method should be followed.
- If the above methods do not work, chemicals which are permitted by the competent authority should be used with the dose allowed.
- If any chemicals used at any stage which are not permitted by the authority, will lead to lose its organic status.

14.3.2 Processing and Manufacturing

All type of processing methods should be done through mechanical, physical or biological method. Fermentation or smoking are examples of this. Use of non-agricultural products, additives, preservatives should be lowered to meet the standard. For packaging purpose, biodegradable and recyclable materials should be used.

14.3.3 Storage and Transport

The quality of the product and its integrity should be maintained throughout the storage and transportation. Organic products should not be mixed with non-organic ones at the time of transportation. Any other products except organic should be kept and handled separately. Storage place and transportation materials should be cleaned with the product and procedure which are permitted by the competent authority. Proper care should be taken to prevent any type of contamination or pest attack.

14.4 PERMITTED SUBSTANCES FOR ORGANIC PRODUCTION SYSTEM

Though organic system uses all eco-friendly measures in every step, still some chemicals are permitted to some extent. But the misuse of permitted chemical has the capacity to alter the ecosystem and hamper the environment. Table 14.1 describes specific chemicals which are permitted to some extent in various stages.

Table 14.1 Permitted substances to be used in organic production by the competent authority

Substances used for fertilization and conditioning	Substances used for pest control	Permitted food additives	Processing aids
Slurry or urine	Plant and animal oils	Citric acid	Ethanol
Straw	Casein	Malic acid	Casein
Wood ash	Natural acids	Ascorbic acid	Gelatine

Contd...

Farm yard and poultry manure	Gelatine	Lactic acid	Talc
Natural phosphate rock	Neem extracts	Calcium carbonates	Beeswax
Sawdust	Seaweeds and their extract	Sulphur dioxide	Egg white
Seaweeds	Mushroom extract	Alginic acid	Kaolin
Magnesium rock	Aspergillus extracted fermented products	Agar	Perlite
Sulphate of potash	Tobacco tea	Sodium alginate	Vegetable oil
By-products of food and textile industry	Extract from Chlorella	Karaya gum	Silicon dioxide
Basic slag	Potassium permanganate	Guar gum	Calcium chloride
Rock potash	Paraffin oil	Locust bean gum	Calcium sulphate
Peat	Sulphur	Mono calcium phosphate	Magnesium chloride
Vermiculite	Copper salts	Ammonium carbonate	Potassium carbonate
Wood charcoal	Burgundy mixture	Potassium carbonate	Activated carbon
Human excreta	Sodium silicate	Oxygen	
Gypsum	Pheromone traps	Argon	
Zeolites	Sterilized insect males	Sodium hydroxide	
By-product of sugar industry	Ethyl alcohol	Calcium chloride	
By-product of oil palm, coconut and cocoa industry	Herbal preparation	Magnesium chloride	
Humus from insect and earthworm	Potassium soap	Nitrogen	
Trace elements	Ayurvedic preparations	Calcium sulphate	
Stone meal	Homeopathic preparations	Potassium chloride	
	Microorganisms	Tocopherols	
	Copper salts	Carageenan	
	Mineral powders		

14.5 LABELLING

Labelling of organic product is mandatory according to the rules and regulations. All information regarding the product should be clear and accurate. The date of manufacture, expiry date, components of product, any chemical residue, price, name and address of person or company, processing procedure, etc., should be clearly shown in the label. After the product withstands all the requirements according to the standard, it will be sold as 'produce of organic agriculture'. If in case of any discrepancy, the company or the individual person will be legally responsible. The following standards should meet accordingly to be regarded as "produce of organic agriculture":

- Water and salt content should not be considered in the percentage calculation of organic ingredients.
- If 95% of ingredients are from organic origin, then it will be labelled as "certified organic".
- Where the products (more than 70% and less than 95%) are from organic origin, it will be considered "organic".
- If less than 70% products are not from organic origin, then it will not be labelled as organic.
- If the content of spices and herbs are less than 2%, it should be shown as spices and herbs without their percentage.
- These products should not be labelled as GE (genetically engineered) or GM (genetically modified) to avoid any type of misunderstanding among customers.

14.6 ACCREDITATION PROCESS

Accreditation regulations are known as accreditation regulations (2001) under National Organic Programme for Organic Production. These regulations are applied to all certified agencies to carry out certification process. Any certified body without proper accreditation from NAB is not considered a certified agency. Government of India has all rights to amend or modify any criteria described before, if it seems inappropriate.

14.7 NATIONAL ACCREDITATION BODY (NAB)

National steering committee also acts as NAB. After proper evaluation, NAB is considered the accreditation body of various inspection and certification agency. This body comprises members from APEDA, Ministry of Agriculture, Ministry of Commerce and Industry, Coffee Board, Spice Board, Tea Board. The committee shall meet as and when required but at least twice per year. The chairman of NSC acts as the head of this committee. This committee conducts surprise field visits, sample analysis, etc.

14.8 APPLICATION FOR ACCREDITATION

An individual body or company seeking accreditation can apply to the authorized office in prescribed format which is available at APEDA with an accreditation application fee of Rs. 15,000 in the first year. The accreditation certificate is valid for a period of 3 years. For the renewal of accreditation certificate, a fee of Rs. 10,000 should be deposited for a period of 3 years. In case of failure to get accreditation certificate, the total fee will be refunded by deducting 25% of processing charge.

All agencies who get the certification and accreditation certificate are allotted with a unique accreditation number. The logo used under the certification process is known as "India Organic". This logo refers to that all the products are organically produced in India. This logo can be used by all accreditation bodies.

14.9 TERMINATION OF ACCREDITATION

The certificate of accreditation is liable to be suspended for a temporary period or permanently if violation of any norms is observed. In case of suspension, the NAB shall have the powers to nominate any other certification agency to continue the work of certification in order to protect the interest of the operators.

14.10 CATEGORIES FOR ACCREDITATION

Accreditation can be granted for all the products given below:

1. Organic agricultural products
2. Organic processing operations
3. Wild products
4. Forestry
5. Organic animal production and processing

14.11 GENERAL PROVISION FOR INSPECTION

Inspection and certification agencies visit the field without any prior notice, prepare their report in a confidential way and proper consistency. They should stick to the regulations prepared by the authority. Any persons related with the owner should be excluded from the agencies. A confidentiality agreement has to be signed by the members to keep all the information obtained at any stage of inspection as confidential.

14.12 CONCLUSION

Growing of any farm produce or product is not the only factor to refer it as organic. For proper marketing of an organic product, certification is needed. Customers should trust this as purely organic and for this reason, NAB is formed to accredit several agencies as a certified body. These certified agencies follow all the standards, do surprise visit, and check the quality to give certification. Therefore, all the regulations should be properly delivered to the farmers by the government through training, workshops or through NGOs. Otherwise, the whole effort will go in vein.

14.13 QUESTIONS

Long Questions

1. What is the process of labelling of a product?
2. Why accreditation is important in organic farming?
3. How to package and transport an organic product? List out all the permitted substances used in organic product?
4. What are the requisite standards to get the logo, "India Organic"?
5. Explain the pest management in organic system at the time of processing.

Short Questions

Write a Short Note on:

1. Accreditation body
2. NAB
3. APEDA
4. "India Organic" logo
5. Food preservatives in organic system
6. Use of fertilizers in organic system
7. Certification of product
8. Application for accreditation
9. Storage of organic product
10. Packaging and transport of organic product

REFERENCE

National Programme for Organic Production, Department of Commerce, Ministry of Commerce and Industry, New Delhi. 6th edition, 2005.

15 CHAPTER Certification Process and Standards of Organic Farming

15.1 INTRODUCTION

Certification process in organic farming is one of the major regulations made by the government. Without organic certification, a grower or individual can't market his product. Selling of organic products by labelling it as organic without proper certification does not make it organic, it is illegal. Thus, this chapter explains the importance and purpose of certification, its standards and how to get it.

15.2 ORGANIC CERTIFICATION

It is a certification process where a farmer, owner or a company who produces organic products gets a certification to market its produce as organic. The products refer to any type of agricultural or wild products, seeds, vegetables, fertilizers, etc. Any seed producer, fertilizer company, restaurants, etc., can apply for organic certification process. But the standards differ from country to country. And all the standards should be followed from growing to selling which are given below:

- Restriction on use of any synthetic chemicals at any stage of farming
- Chemical-free land for more than 2 or 3 years
- Maintaining detailed sales and production record for audit
- Keeping organic and non-organic products separately from harvesting to selling

15.3 PURPOSE OF CERTIFICATION

Organic certification is needed to preserve its quality and any type of fraud which is increasing parallelly with the demand of organic products. Certification facilitates and regulates the marketing of organic product through different certified agencies. These agencies do surprise inspection, audit, quality check considering the minimum requirements of organic standards by the national government. After certification, the individual or company can use the logo "Certified

Organic" which serves as a product assurance. In organic system, certification stands the product and producer as organic.

15.4 CERTIFICATION PROCESSES

For the certification process, the owner or company have to go through a number activities which are described below.

- **Study:** The producer should study the entire manual of organic production and its standards to get a clear idea on what is allowed and what is not.
- **Compliance:** The owner should modify some of his farm structure or any method to comply with the standards according to the guidelines.
- **Documentation:** Proper documentation needed which should contain farm history, current situation, soil test report and water test report.
- **Planning:** A detailed production plan of one year should be submitted and it should carry the information regarding seed, fertilizer, pest control method, harvesting techniques, storage and transport, selling, etc.
- **Inspection:** An annual on-farm inspection and an oral interview is required. Examination reports should be kept properly.
- **Fee:** The producer should deposit the required fee for annual surveillance and to get the certification as a mark of its quality.
- **Record-keeping:** A detailed written record must be maintained by the producer containing all information about all activities and techniques followed on daily basis. The record should be presented before the certifying agency at the time of surprise visit.

15.5 CERTIFICATION AND PRODUCT LABELLING

Certification makes the consumers to believes that the product is organic and avoids any type of fraud. Labelling the product as organic increases its demand in the open market. The organic products can get three types of certification. If the entire product and its ingredient is organically produced, then it can be labelled as "100% organic" (Fig. 15.1). If more than 95% of the product as well as ingredients are of organic origin, then it can be labelled as "organic". If the product and 70% of its ingredients are of organic origin, then it can show its level as "Made From Organic Ingredients". If less than 70% of its ingredients are organic origin, then this product cannot display its label as organic, but it can be shown in its components.

Fig. 15.1 Logo for organic products in India.

15.6 CERTIFICATION AROUND THE WORLD

Different countries have different organic rules and regulations (Fig 15.2). The United States of America, Europe, Japan and Germany have comprehensive rules and regulations. Countries where there is no rules and standards, the certification process is handled by non-governmental organizations and private sectors.

In China, China Green Food Development Center functions as the certification body. it has 2 grades, A and AA. Grade A refers to use of synthetic chemicals to some extent while AA is very strict and does not allow use of any synthetic chemicals.

In India Agricultural Food Processing Export Development Authority under Department of Commerce, is the controlling and competent authority for the certification of the agricultural products. 11 certification agencies are identified in India which function under NPOP. In 2006, it was announced that the standards of certification matched equivalently with European union. USDA NOP has also given its conformity on India's NPOP regulations.

Japan	Malaysia	Organic Alliance Malaysia	NOP-USA	EU	P.R. China
USA	USA	Control Union Holland	USA	USA	USA
Germany	Germany	Brazil	Germany	Germany	Germany

Fig. 15.2 Certified logos of some countries.

15.7 INSPECTION AND CERTIFICATION AGENCY

The national accreditation body is the regulatory authority to accredit an agency for inspection and certification of the organic producer. After meeting all the standards and regulations, Agricultural and Processed Food Export Development Authority (APEDA) gives certification to that agency for further inspection and certification of individual producer or company. The accreditation certificate by APEDA contains the following information in the certificate:

1. The certification accreditation number
2. Name and address of the certification agency
3. Types of the work to be covered by the agency
4. Date of issue and date of expiry

15.8 FUNCTION OF INSPECTION AND CERTIFICATION AGENCY

The inspection and certification agency prepares an annual report containing the annual turnover about its product and number of certificates given to the owner or producer, etc. This report also contains new methods adopted by the agency, any changes in the office staff, about the licence operators, locations covered and any infringement or fraud case detected during the inspection. The evaluation committee reviews the annual report and sends it to NAB through proper channel for hearing of the decision. The agency suggests any measures and actions to be taken by the APEDA. The NAB announces the final decision by observing the report. The following actions can be taken by NAB:

- Renew of date, if period expires.
- Extension of authorization period for re-evaluation.
- Enforcement of new standards required for corrective action within a given time.

The inspector assigned to inspect an operator or company should follow all standards classified in the manual. Licensed operator does not have any right to choose or recommend an inspector. But in case any problem, he can change the certification agency by informing the APEDA. A single inspector should not inspect the same license operator continuously. The agency should inform about all actions taken on the licensed product regularly to the NAB. If found any irregularity and fraud; further action will be taken by NAB.

15.9 PROCEDURE OF CERTIFICATION

Inspection visit: Inspectors of certification and inspection agency go for field visit at least once annually and sometimes do surprise visit. All the information on methods and techniques, inputs, plans, product specification, any earlier infringement or any disciplinary measures should be made available to the inspector by the licence operator. If the produce is growing same type of product as non-organic, inspector can visit the non-organic production unit. The reports made after inspection should be elaborate and clean and should give information about:

1. Product and its specification
2. Date and time of inspection
3. Persons interviewed
4. Documents checked
5. Field facilities visited
6. Inspector's observation
7. Evaluation of compliance and standard check

The method and frequency of visit mainly depend on the type and size of production, outcome of earlier inspections, any complaint received against the producer, contamination and drift risk, etc. A minimum of one visit should be carried out once a year. Surprise visits should also be done as per the minimum required percentage. Timing of inspection should be unpredictable and in irregular intervals.

15.10 ANALYSIS OF THE PRODUCT AND RESIDUE TESTING

Residue testing is one of the major steps to check the presence of any chemicals like pesticides, heavy metals or genetic testing and some other analysis. This testing should be carried out in laboratories which are approved as per ISO 17025 and have proper lab equipment and facilities for testing. The inspection and certification agencies have their own procedures and policies which must include:

- Identification of sample and in which it should be taken for analysis as per the meeting standards
- Indication of collection of any random samples
- Inform to inspectors about requirement of samples and methods.
- Post-sampling procedures
- Fixation of responsibility for sampling payment

15.11 CERTIFICATION DECISION

The final certification decision is taken by the certification agency rather than the persons involved in inspection. Certification is not only limited to the licensed producer, but also needs approval of his products. In case of any exception, it should be clearly mentioned in the report like reason, limited time for exception. The process of certification is as follow:

- All documents related to procedure, processing to final certification should be maintained properly.
- Procedure of extension and updating of certification should be included in the report.
- Certification decision should be clearly mentioned and informed to the producer.
- If denial of certification, reasons should be mentioned clearly.
- All documents should be handed over to another inspection and certification agency, in case the producer applies for it.

The certificate of registration shall include:

- Name and address of licensed operator
- Name and address of certification and inspection agencies
- Reference to the applicable standard
- Product and product information
- Date of issue and validity

15.12 GRANT OF CERTIFICATION

- After completing all inspection, evaluation, if the authorized body is satisfied then the certificate will be issued to the licensed operator.
- The operator can use the certification or grade granted by the agency in his packaging or at the time of advertising as described in the norms.
- In case of withdrawal of certificate, the licence shall be returned to the inspection and certification agencies.
- In case of cancellation of certificate due to violation of rules, infringement or furnishing of false information, the operator cannot apply again for a period of 6 months from the date of cancellation.
- Licence can be granted for one year with the self-declaration of the licensed operator.
- The licence can be altered by the agency by giving one-month notice period to the operator.

- The licence will expire in a given period of time and can be applied for renewal.
- The required fee should be deposited by the operator for applying certification or renewal of certification.

15.13 CONCLUSION

Certification is an unavoidable major step to be followed by the operator to market his product as organic. After fulfilling all the criteria and standards, a certification will be issued to the operator and he can use it for advertising the product. The inspection and certification agency should properly monitor all procedures, techniques and reports before announcing final decision. In case of expiry, it can be applied for renewal. Certification is not only a key factor for marketing but it also increases the market value by advertising as organic as per the norms. Proper measures like workshop, training manuals should be distributed to farmers and operators to gain the idea about certification process.

15.14 QUESTIONS

Long Questions

1. What is NAB? How does it work?
2. How certification process is carried out by inspection and certification agencies?
3. What is the need of certification in organic system? What if, certification process will not be there in the organic system?
4. Explain the extra steps to be taken by an operator to get a certification?
5. How do he inspection and certification agency function?

Short Questions

Write a Short Note on:

1. NAB
2. APEDA
3. NPOP
4. Inspection and certification agencies
5. Evaluation committee
6. Grant of certification
7. Cancellation of certification
8. Organic certification

9. Organic certification in China
10. Certification and labelling

REFERENCES

National Programme for Organic Production, Department of Commerce, Ministry of Commerce and Industry, New Delhi. 6th edition, 2005.

APEDA (Agricultural and processed food products export development authority). (http://www.apeda.gov.in/apedawebsite/organic/Organic_Products.htm#:~:text=The%20APEDA%2C%20Ministry%20of%20Commerce,organic%20farming%20and%20marketing%20etc.)

Yadav, A.K. Training Manual Certification and Inspection Systems in Organic Farming in India. National Center of Organic Farming, Government of India, Ministry of Agriculture, Department of Agriculture and Cooperation. 1-67.

16 CHAPTER

Marketing and Export Potential of Organic Products

16.1 INTRODUCTION

The trend of organic agriculture is increasing day by day. In the coming years, it is expected that the trade and marketing of organic products will increase from 10 to 15% to 25 to 30%. Marketing of organic products has made it possible for the farmers to get high returns. However, the trade and export of organic products are still not well known to the farmers of developing countries as compared to developed countries. In India most of the metropolitan cities like Delhi, Mumbai, Kolkata, Chennai, Bangalore, Hyderabad, Bhubaneswar sell the fresh organic products. Export of organic products is mainly done with Middle East, Asia, Africa, Europe, etc. The organic product business is one of the fastest growing business as in recent days. Now, it is the responsibility of the government and non-government organizations, Krishi Vigyan Kendras, extension workers to create awareness among the farmers for production and trade of organic products. This chapter explains about trade of organic products, export of organic products and marketing in India.

16.2 CURRENT STATUS OF ORGANIC PRODUCTS IN THE WORLD

According to BioFach 2010, the organic market of the world is now managed by 1.4 million farmers and 35 million ha land. Countries having largest area of organic farm is Oceania, Europe, as Latin America. Countries having highest number of organic producers are India, Uganda and Mexico. Nearly 141 countries are growing organic food while Australia has maximum land under organic farming. India is the largest producer of organic cotton. China is the largest exporter of organic beef while Israel is the largest consumer of organic egg.

16.3 TRADE OF ORGANIC PRODUCTS IN INDIA

Most people in India believe that the organic products are only for developed countries and the products produced in India is only for the purpose of export. But the fact is that more than 60% of parents prefer organic foods compared to non-organic foods. Nearly 50% of organic food is targeted for the export while other 50% is for consumption in India. But most of the farmers or producers are producing organic foods for their own consumption to avoid any type of health risk. Though organic food prices much higher than non-organic foods, still people want to pay high price considering the health benefits. The demand of organic products is now increasing not only in the metro cities but also in rural India. Only the differences are rural producers grow organic food for their family while in big cities we can find an organic store in every big shopping mall. Although the domestic consumption of organic products in India is low but India stands in the list of leading suppliers of organic spices, herbs and basmati rice.

According to the data released on 31st March, 2020, the total certified area under organic system is 3.67 million hectare and the production is 2.75 million tonnes. Madhya Pradesh has the highest area under organic farming followed by Rajasthan and Maharashtra. In 2016, Sikkim has converted its total cultivable area under organic production. As per the data by FIBL and IFOAM year book 2020, India has gained 8th place in terms of world organic agricultural land and 1st in terms of total number of producers. The list of potential organic products which are exported to foreign countries are mainly:

1. Organic meat
2. Organic dairy
3. Organic beverages
4. Organic textiles
5. Organic cosmetics
6. Organic fruits and vegetables
7. Organic tea
8. Organic coffee
9. Livestock
10. Organic spices

16.4 SIGNIFICANCE OF ORGANIC FARMING IN INDIA

In India green revolution has developed the agriculture but the negative effect decreased the soil fertility as well as increased the input cost. Therefore, organic system can be a way towards profitable farming where less input is used to get high yield. In organic system there is less or no chemical use and locally available materials are used in place of chemicals.

Organic products cost higher price than non-organic products. Thus, growing of organic products can return high profit to the Indian farmers.

16.5 DEMAND OF ORGANIC PRODUCTS IN INDIA

In recent reports it was observed that the consumption and demand of organic products in India is rarely 10%. Supermarkets, wholesalers or restaurants are the major channels for delivering the organic products from farm to consumers' plate. The products available in domestic market are mainly rice, wheat, coffee, tea, fruits, vegetables, pulses, etc. Small farmers sell their products mainly through wholesalers. Large producers market their products through supermarkets or restaurants.

The organic market break-up has been done according to different regions like east India, south India, central, west and north India. These products are categorized to organic spices, organic beverages, organic processed foods, organic vegetables, organic meet and dairy, organic cereals and organic fruits. The organic products are mainly distributed and sold in supermarkets or hypermarkets, convenience stores, online, etc. Some of the major players in Indian market are:

1. Mother Earth
2. Nature Bio-Foods Ltd.
3. Suminter India Organics Pvt. Ltd.
4. EcoFarms (India) Ltd.
5. Phalada Agro Research Foundations Pvt. Ltd.
6. Organic India Pvt. Ltd
7. Conscious Food Pvt. Ltd.
8. Nourish Organic Foods Pvt. Ltd
9. ElWorld Agro
10. Sresta Natural Bioproducts Pvt. Ltd.

16.6 DEMAND OF ORGANIC PRODUCTS IN EXPORTING COUNTRIES

The importance of organic products is well known in developed countries compared to developing countries. Therefore, more than 50% of organic products are exported to the developed countries like America, Europe, France, Asia, Africa, Australia. The products exported to these countries are mainly herbs, spices and basmati rice, fruits, vegetables, coffee, etc. For exporting of the products, except tea, the exporters procure organic products from the small and marginal farmers and export through export companies. But for export of tea, the tea estates are there which directly export it.

As of the data released by APEDA (2019-2020), 6.389 million tonnes is exported with total value of INR 4686 crore. Table 16.1 and Fig. 16.1 shows the products and amounts exported in 2019-2020.

Table 16.1 Organic items and the percentage of export

S. No.	Items	Percentage of export
1.	Soya meal	45.87%
2.	Oilseeds	13.25%
3.	Tea and coffee	9.61%
4.	Cereals and millets	8.19%
5.	Spices condiments	5.20%
6.	Dryfruits	4.98%
7.	Sugar	3.91%
8.	Medicinal plants	3.84%
9.	Others	5.15%

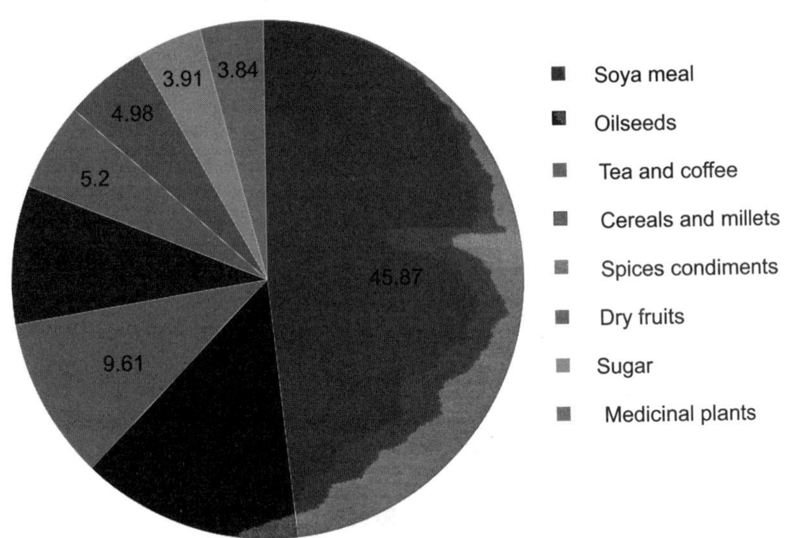

Fig. 16.1 Organic products and its percentage of export.

16.7 CHALLENGES IN ORGANIC FARMING

In recent days, organic farm revolution has gained more attention as compared to green revolution. One major drawback is lack of knowledge in organic farming system in most of the rural areas in India. Organic farming system and products have impressed many people of large cities but not yet gained interest in rural areas. Due to less knowledge and clarification, the farmers produce organic products only for their own consumption. Though initial cost in organic production system is slightly higher than conventional method, but the profit always stands above in the organic method. Therefore, the biggest challenge for the government is to deliver proper knowledge and awareness about organic farming system and its high profit with the help of non-government or government organizations, KVKs, workshops, etc. Table 16.2 explains the challenges and solutions in import and export of organic products.

Table 16.2 Problems and solutions for import of organic products in India

Sl No.	Problem	Solution
1.	Complicated documentation and paper work which is time consuming	Need a fast track for export
2.	Poor customer service by Indian exporters	Proper client follow-up system and increase of proper service quality
3.	Lack of proper information regarding certification and certified organic products	Need of a website containing all information regarding the product and more promotion activities
4.	Low quality due to contamination	Proper processing and post-harvest management
5.	High price	Need of realistic price
6.	Slow mode of payment	Bank system needs higher standards of payment system.

16.8 CONCLUSION

Though India stands among the top leaders in export and production of organic products, still it needs more attention to develop more interest for more production and use in our country. The traders, government and estates have major role in the marketing of organic products. More promotion and advertisement of organic products can result in the vertical growth of farming as well as reduction of health risk of people. The certification process and information regrading this must reach to rural villagers so that they can adapt this system. Implementation of flexible regulations, consumer-oriented follow-up, easy payment methods, transparent marketing, etc., can develop the organic production market in near future in India.

16.9 QUESTIONS

Long Questions

1. Write the global status of organic products. How it can be improved?
2. How the organic products are exported? What are the problems faced by the producers in this system?
3. Why organic products have not impressed rural people compared to urban area? Explain.
4. Explain the significance and opportunities of organic products in India.
5. Describe the potential organic products to export. How India has reflected its existence in the world in terms of export of organic products?
6. What is the structure of organic food industry in India? Explain.
7. What are the future perspectives of Indian organic food market? How will it perform in upcoming days? Explain.
8. What is the breakup of the organic food market in India on the basis of distribution channel, product type?

Short Questions

Write a Short Note on:

1. APEDA
2. NPOP
3. Organic tea
4. Demand of organic products in India
5. Organic spices
6. Basmati rice
7. Significance of organic products
8. "India Organic"
9. IFOAM
10. Role of KVKs in organic production system

REFERENCES

APEDA. Accredited inspection and certification bodies. In National Programme for Organic Agriculture (NPOP)-standards, 2020.

Cornish PS, Stewart TEB. Certification – Case studies with Australian market gardeners. In: Thompson, R. (ed.) Cultivating Communities.

Proceedings of the 14th IFOAM Organic World Congress, 21–28 August 2002, Victoria, Canada. Canadian Organic Growers, Ottawa. 2002, 222.

Singh S. (2009) Organic produce supply chains in India: Organization and Governance, CMA publication no. 222, IIM, Ahmedabad.

Subhash C.R. (2004) Data Envelopment Analysis: Theory and Techniques for Economics and Operations Research, Cambridge University Press.

The Economics of Organic farming: An International Perspective (eds) (1994) by Lampkin and Padel, CAB International.

Unit 6

Practical

17 CHAPTER

Visit of Organic Farms to Study the Various Components and Their Utilization

17.1 INTRODUCTION

An organic farm contains several components like livestock, trees, compost pit, store house, farmhouse which are related with each other. To get rid of any type of contaminants in an organic farm, a buffer zone should be there in an appropriate location (Meena et al., 2013). An organic farm follows all the prescribed norms and standards to produce quality materials. The objectives for organic field visit are given below.

- To know the layout and components of organic farming.
- To acquaint with all the implements and livestock of the farm.
- To gain knowledge about all the procedure followed in the farm.
- To know the problems faced in organic farming.
- To identify the opportunities for organic produce.

17.2 PRINCIPLE

The basic principle behind the field visit is 'seeing is believing'. The information and knowledge collected directly from the owner or farmer of the farm can help in motivating in a more practical way (Meena et al., 2013). While visiting the farm, more focus should be given on the below points:

- The technology and practical procedure
- Condition of plants as well as livestock
- Contamination and how to control it
- Compost pit and its proper use

17.3 REQUIREMENTS

1. Organic farm
2. Field notebook
3. Camera
4. Vehicle

17.4 PROCEDURE

- Take an appointment from the farmer or the owner to know his availability.
- Make a proper planning like about the location of farm, how to visit, convenient time for both as well as your purpose of visit.
- Be punctual and visit the farm according to the convenient time.
- Collect all information regarding the farm, its components, business plan, etc., and note it in your field record.
- Don't pluck anything without asking the farmer or owner and maintain proper discipline.
- Praise and greet the farmer for his good work and quality time given.

17.5 COMPONENTS OF ORGANIC FARMING (FIG. 17.1)

1. **Organic manures:** Organic manures like FYM, compost or vermicompost, green manure, cover crop, crop residue management, biogas slurry is the potential ways to increase the nutrients as well as soil health. Fish manures, seaweeds and some licensed fertilizers like rock phosphate can also be used.

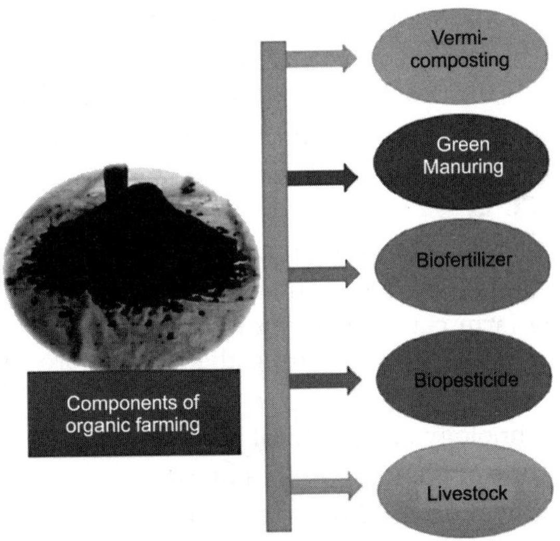

Fig. 17.1 Components of organic farming.

2. **Biological pest control:** Botanical pesticides, biocontrol agents are the best practices to control pests in organic farming. The growth of natural enemies can also be encouraged here.
3. **Weed management in organic way:** Botanical or mycoherbicides are the best alternative for chemical herbicides. Bioagents can

also be employed for weed management. Mechanical weed control or roughing can also be followed.

4. **Organic nutrient management:** For a balanced nutrient management in an organic farm, a farmer can follow crop rotation, use of crop residue, green manuring, intercropping with legumes, biofertilizer, organic manure, etc. (Clark et al., 1999).

17.6 OBSERVATION

1. Name of the farmer:
2. Address:
3. Phone number:
4. Area of the farm:
5. Details about the land (owned/ lease) (wetland/garden land/ dry land):
6. Soil type:
7. Experience in organic farming:
8. Irrigation source:
9. Farm tools, implements and machinery: (Table 17.1)

S. No.	Name	Number
1.		
2.		
3.		

10. Table 17.2: Livestock

S. No.	Name	Number
1.		
2.		
3.		

11. Table 17.3: Cropping

S. No.	Crop	Variety	Area	Yield
1.				
2.				
3.				

12. Table 17.4: Details of organic farm (Previous year)

Income	Expenditure	Net profit	Remarks

13. Table 17.5: Problems faced and suggestions

S. No.	Name of field	Problems faced	Suggestion by farmer
1.			
2.			
3.			

14. Sketch of physical layout of organic farming:

17.7 DON'TS

- Be punctual.
- Don't be afraid to clarify your doubts.
- Maintain discipline and don't enter any unauthorized area without permission.
- Take water bottle, food item and umbrella with you.
- Interact with the farmer regarding the farm issues only.
- Don't pluck anything or disturb animals.

REFERENCES

Clark MS, Horwath WR, Shennan C, Scow KM, Lantini WT and Ferris H. 1999. Nitrogen, weeds and water as yield-limiting factors in conventional, low-input, and organic tomato systems. *Agriculture Ecosystems & Environment*, 73: 257-270.

Meena, R.P., Meena, H.P., Meena, R.S. 2013. Organic farming: concept and components. *Popular Kheti*, 1(4): 5-14.

18 CHAPTER Preparation of Enrich Compost, Vermicompost, Bio-fertilizers/Bio-inoculants and Their Quality Analysis

18.1 PREPARATION OF ENRICHED COMPOST

Waste mica and rock phosphate are two most potential products which can be prepared through pit or trench method (Biswas 2008).

18.1.1 Procedure

1. Fill the pit with biodegradable organic matter like agriculture and animal waste, farm waste, etc., in 5 to 6 layers and each layer should be 20 cm thick.
2. Spread the rock phosphate and then waste mica in the pit in layer wise.
3. Add cow dung slurry through sprinkling over the layers in pit.
4. Moisture content of 60% should be maintained in the pit.
5. Turning should be done on interval basis for proper aeration.
6. The surface of the pit is sealed with cow dung and after that cover the pit with any type of covering material like polythene to protect any type of damage.
7. Leave it for 4 months to increase the nutrient content, .i.e., high P_2O_5 and K_2O.

18.1.2 Economics

Cost of 1 kg P_2O_5 in DAP = 16.22 (Rs)

Cost of 1 kg K_2O in KCl = 7.43 (Rs)

Cost of preparation of enriched compost of waste mica and rock phosphate is = 7 (Rs)

Therefore, the use of enriched compost and waste mica is much better than using DAP and KCl.

18.1.3 Advantages

- It is cost effective and eco-friendly.
- It contains higher amount of nutrient then other chemical fertilizers.

- It also properly maintains the C:N ratio.
- Increase of available N, P and K in the field.
- More crop residue can be recycled back.

18.2 PREPARATION OF VERMICOMPOST

18.2.1 Materials Required

- Cemented tank
- Cow dung
- Plastic net
- Earthworms
- Gunny bags
- Thatched roof
- Water
- Waste materials
- Polythene sheet

18.2.2 Procedure

1. **Selection of site:** It can be prepared in any place such as open area or unused buildings or cattle shed. But if it is in open area, then it must be shady or have thatched roof.
2. **Selection of earthworm:** The earthworms having high digestion power to eat biomass or excreta are normally preferable. The locally available earthworms can also be used but these are less potential. The most suitable are red worms (*Eisenia foetida*) and African earthworms (*Eudrillus engenae*).
3. **Waste material collection:** Biodegradable waste such as cattle dung, animal excreta (goat, pig, poultry etc), agriculture waste, crop residue, kitchen waste can be used. All waste should be pre-digested with cow dung and cow dung is also required to be decomposed properly before used for vermicomposting. Heavy metals and non-biodegradable waste should not be used.
4. **Preparation:** Construct a cement tank according to the availability of waste materials. The waste materials should be kept under sun in a heap and cow dung slurry should be sprinkled for fast decomposition. In the first layer, place a one- to two-inch layer of soil or sand and then spread the pre-decomposed waste materials. Hard materials can be chopped for easy decomposition. After placing the waste up to one feet height, earthworms is released .i.e. 1000-2000 worms/m^2. Cover it with polythene or gunny bags. Sprinkle water to maintain moisture content of 70-80%. Protect it from rain or direct

sunlight. Water sprinkling is not required when 90% of biomass is decomposed. Formation of granular structures confirms the maturity of biomass. Scrap the upper portion of the pit to collect the vermicompost. Preparation of vermicompost is represented in Fig. 18.1.

5. **Separation:** Harvested vermicompost should be kept under shade for separation of worm for 6 to 12 hr under shade. Separate the earthworm and dry the vermicompost to hold moisture of 20% for future use. Otherwise, sieve it and pack in a proper packet.

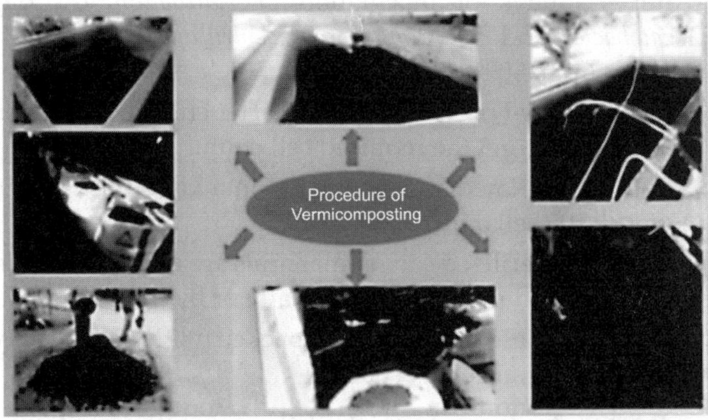

Fig. 18.1 Procedure of vermicomposting.

18.3 PREPARATION OF BIOFERTILIZER

18.3.1 Media Required for Mass Culturing of Biofertilizer

- *Azatobacter:* Waksmanna No 77 broth
- *Azospirillum:* Dobereiners mallic acid broth with NaCl
- *Pseudomonas:* Kings B broth
- *Phosphobacteria:* Pikovaskys broth
- *Rhizobium:* Yeast extract mannitol agar
- *Trichoderma:* Potato dextrose broth

18.3.2 Carrier Material Required

- Charcoal
- Coconut shell
- Peat
- Press mud
- Rice husk
- Leaf manure

18.3.3 Mass Production

- Isolate the microorganism from the soil or root and go for serial dilution technique.
- Test the microorganism in lab conditions as well as in field to know their potentiality.
- If it is effective, go for refinement of inoculum and start mass production.
- Isolated microbial culture was subcultured in required nutrient media and then kept in rotary shaker at 30±2°C for their growth.
- After 4-5 days of inoculation of *Rhizobium*, 6-7 days for *Azatobacter* and 5-7 days for *Azospirillum*, the colony count reaches up to 10^{10} to 10^{11}.
- For commercial production, this starter culture is transferred to a fermenter to get the required cell count.
- After that, the media is inoculated in a large flask to get a cell count of 109 cells/mL.
- For selling, sterilized carrier materials are mixed with microbial inoculants and mixed well manually or by machine.
- After that, these are properly packed with sealed polythene bags sell (Curnoe et al., 2006).

18.3.4 Quality Analysis of Enriched Compost, Biofertilizer and Vermicomposting

(a) Physical Parameters

1. **pH:** Take 30 gm sample after separating the granular parts through 2 mm sieve and mix in 60 mL distilled water. Stir it properly and measure the pH by using digital pH meter and record the data.
2. **Electrical conductivity (EC):** The amount of total soluble salts present in a sample is known as electrical conductivity (EC). It is expressed as millisimens/cm (mS/cm). Mix 5 gm of sample in 50 mL of distilled water and stir it properly. Leave this sample for 8 hr to settle the suspension. After that, dip the electrode of conductivity cell to measure the EC and record it (Singh and Sharma 2002).

(b) Chemical parameters

1. **Estimation of organic carbon:** It can be measured by the empirical method. Take 1 gm of fine oven, dried sample in a silica crucible and heat it by a muffle furnace at a temperature of 550°C for 2 hr. Leave it to cool in a desiccator and take the weight again. The formula to measure the organic carbon (%) is given as:

Organic matter (%) = (Initial mass – Final mass) / Initial mass × 100

Organic carbon (%) = Organic matter (%) / A. where, A is a constant = 1.724

2. **Estimation of nitrogen (N):** It can be estimated by following Kjeldahl method which involves 2 steps, i.e., conversion of N to NH^{4+} and the determination of NH^{4+} in digested sample.

- For digestion of the test sample, take 0.5 gm dry sample with 20 mL sulphuric salicylic acid and shake gently so that the sample comes in contact with the reagent. Leave it for overnight and then heat the mixture gently for 5 min after adding 5 gm sodium thiosulphate. Avoid frothing at the time of heating. After cooling, add 10 gm sulphate mixture to that and digest for 1 hr. To avoid bumping, glass beads can be used. After cooling, go for distillation of the sample.
- After cooling of digested sample, transfer 10 mL to a vacuum jacket of micro-Kjeldahl apparatus. Take a conical flask with 10 mL of 4% boric acid solution with bromocresol green and methyl red indicator so that the outlet of condenser can be dipped in it. Add aliquot digest and wash funnel apparatus with 10 mL boric acid and 2-3 mL deionized water. When distillation is complete, titrate boric acid against N/200 H_2SO_4 and maintain control side by side at same end point.

N content (%) =

$$\frac{TV \times 0.00007 \times 100 \times 100}{0.5 \times 0.5} (1mL \text{ of } \frac{N}{10} H_2SO_4 + = 0.000014 \text{gm N})$$

where sample weight = 0.5 gm

Volume of digestion = 100 mL

Normality of H_2SO_4 = N/200

Titrant value (TV) = Sample TV-Blank TV

Aliquot taken = 5 mL

3. **Estimation of phosphorus (P):** Pipette out aliquot from sample digestion to a volumetric flask. Add 10 mL vanadomolybdate reagent. Make the volume by adding deionized water and shake it. Observe the yellow colour and take the absorbance at 420 nm with a spectrophotometer. Determine the phosphorus concentration by using prepared standard curve.

P(%) = Sample concentration (ppm) ×

$$\frac{1}{\text{Weight of sample}} \times \frac{100}{\text{Aliquot (mL)}} \times \frac{\text{Final volume (mL)}}{1000}$$

4. **Estimation of potassium (K):** It can be estimated by the flame photometric method. The sample was atomized and then note the readings. The K concentration can be determined by preparing a standard curve and then multiply with dilution factor.

5. **Estimation of sulphur (S):** It can be estimated by the sulphate turbidimetry method or wet ashing of plant sample. Take 10 mL aliquot from diacid digestion (a described in phosphorus estimation) in a volumetric flask and add 25 mL salt buffer solution. Adjust the volume up to 50 mL by adding deionized water and mix thoroughly. Pipette out 1 mL solution in to a conical flask. Add 1 mL of 0.05% gum acacia, 6N HCl and dissolve 0.5 gm of $BaCl_2$ crystals. Take the absorbance in spectrophotometry in 420 nm and estimate the sulphur content by prepared standard curve (Griffths et al., 2007).

$$S\ (\%) = \frac{R}{\text{Sample} \times 100}$$

6. **Estimation of magnesium (Mg) and calcium (Ca):** Take an aliquot sample containing Mg of 3.0 mg and dilute it to 10 mL. Add 10 drops of potassium cyanide, hydroxylamine hydrochloride, potassium hexacyofarrate and triethanolamine solution. For Mg determination, heat the solution for 3 min and add 10 drops of Erichrome Black-T (EBT) indicator solution after cooling. Titrate it with EDTA. For Ca determination, add 1 mL calcon solution and 2.5 mL NaOH. Titrate it against 0.01N EDTA by using EBT indicator till the colour changes to blue.

Ca or (Ca + Mg) meq/gm =

$$\frac{(\text{mL of EDTA consumed} \times \text{Normality of EDTA})}{\text{Aliquot taken}} \times \text{Volume made up}$$

REFERENCES

Ali M, Griffiths AJ, Williams KP, Jones DL. 2007. Evaluating the growth characteristics of lettuce in vermicompost and green waste compost. *European Journal of Soil Biology*, 43: S316-S319.

Biswas, D.R. 2008. Preparation of enriched compost. ICAR: *A Science and Technology Newslette,* 14(3): 1-2.

Curnoe WE, Irving DC, Dow CB, Velema G, Unc A (2006) Effect of spring application of a paper mill soil conditioner on corn yield. *Agronomy Journal,* 98: 423–429.

Singh, A., Sharma, S. 2002. Composting of a crop residue through treatment with and microorganism subsequent vermicomposting. *Bioresource Technology,* 85(2): 107-111.

CHAPTER 19

Indigenous Technology Knowledge (ITK) for Nutrient, Insect, Pest Disease and Weed Management

19.1 INTRODUCTION

Farmers are practising on weed, disease and insect control by using the locally available material from the starting of agriculture. The uses of indigenous technology knowledge (ITK) are mentioned in Vedas and Upanishads. Indigenous technology knowledge (ITK) can be referred to as the use of locally available natural resources by the farmers to get rid of the weeds, diseases and insect pest (Anbu et al., 2018). According to World Bank (1998), the ITKs must be:

- Local
- Transmitted orally
- Learned through repetition
- Tacit knowledge
- More experiential and less theoretical
- Constantly changing

19.2 CHARACTERISTICS OF ITKs (Pandey et al., 2017)

- Qualitative in nature.
- Need a holistic approach.
- Dynamic.
- Learnt through observation and experience.
- Reflects local practice.
- Proper use of locally available natural resources.

19.3 SOURCE OF ITKs

- Local farmers
- NGOs
- Community heads
- Old people of the locality
- Extension offices
- Old records
- Songs and poetry

19.4 SOME ITKs FOR DISEASE, INSECT PEST AND WEED MANAGEMENT

- Soaking of paddy seeds in milk can reduce the incidence of rice tungro virus and stunt virus disease.
- Soaking of paddy seeds in cow urine and acorus calamus mixed in the boiled water after cooling. The floating seeds should be removed and the dipped seeds used for sowing. This reduces the incidence of seed-borne diseases.
- Sundrying of cow urine treated sorghum seeds can control head smut disease.
- Dusting of dried neem fruits powder at 200 kg/ha can control stem borer and fungal disease.
- Soaking of maize seeds in warm water for 3-6 hours can be effective in control of shoot borer incidence.
- Ash is dusted in the pearl millet fields can reduce the incidence of green leaf hopper.
- Pouring of neem cake extract near the shoot region of sorghum can reduce the shoot borer attack.
- Red gram seeds coated with red earth slurry followed by drying can reduce the incidence of storage insects.
- Corriander and sorghum mixed cropping can reduce the incidence of witch weed.
- Dipping of banana suckers in neem oil and water solution (1 litre neem oil/100 litre of water) can reduce the rhizome rot incidence.
- Intercropping of sunflower in between mango plants can increase honeybee population which ultimately increases pollination.
- To get rid of damping off disease in nursery, tomato seeds are soaked with the leaf extract of bougainvillea for 6 hours.
- Growing onion or garlic as a border crop of tomato prevents from fruit borer attack.
- A spray solution of 8 crushed leaves of aloe vera plant with one litre of water can prevent the fruit rot disease in brinjal.
- An equal proportion of ash and turmeric powder can be dusted to control aphid in brinjal field.
- A mixture of tobacco waste and neem cake extract can be supplied in field to get rid of thrips and stem borer.
- A solution of milk and water in 1:20 ratio when sprayed in tobacco filed, it reduces the incidence of tobacco mosaic virus.
- Application of lime can reduce the coffee wilt in orchard.

- Insect pest population in rice field can be minimized by application of leaves or seeds of custard apple by broadcasting method.
- To prevent the attack of BPH, rice bran, kerosene and molasses or jaggery in 7:1.5:1.5 ratio mixed properly and placed near BPH infected plant which act as a BPH repellent.
- Fresh cow dung application near the collar region of chili plant can prevent damping off disease.
- Leaves of acacia plant can be used in water channels to control brown spot disease.
- Mixing of dried neem leaf powder at 1% can control stored grain pests.
- Garlic and crushed green chili in equal quantity with 200 litres of water can control jassid and aphid attack (Shailaja et al., 2012).
- A mixture of whey (2 litres), tobacco powder (200 gm), *Aloe barbadensis* (2 leaves) with 15 litres of water is kept undisturbed for 15 days. After that, it is filtered and the filtered solution spray can effectively control pod borer in gram crop.
- In the rice growing regions of Himachal Pradesh, farmers sweep the field with the Nirgundi (*Vitex negundo)* branches to reduce the insect population.
- In Rajasthan and Arunachal Pradesh, several traditional storage structures like stone bins, mud bins or bamboo bins are used for storing of food grains to prevent the attack from insect pest.
- A solution of water (1 litre), tulsi (100 g) and soap solution of 1 mL can effectively used to control wide range of insects like bug, aphid, mosquito etc.
- Application of botanicals mixture of parthenium, segun and akand leave powder extract at 100 mL per litre of water can reduce the weed incidence.
- Inclusion of more legume crops in crop rotation can also effectively control the disease incidence.

19.5 SOME ITKs FOR NUTRIENT MANAGEMENT

- To recover the soil fertility, FYM or sheep penning can be done in the field.
- Groundnut cake can be applied in the field to increase the yield.
- Coconut milk solution of 5% concentration can be used as a growth booster.
- Use of animal urine, sheep manure, wood ash, cow dung, camel feacal in field can increase the soil fertility.

- Growing of certain local weeds like fog (*Calligonum polygonoids*), bordi (*Zizyphus mauritiana*), bui (*Kochia indica*) can also increase the nutrient content in soil (Shreth et al., 2009).
- Use of ants, frogs and earthworms can increase the soil health and soil physical property.
- Zal technique, one traditional method, a combined irrigation and application of organic amendments in small pits can increase the nutrient and water retention capacity.
- To increase the germination process, radish seeds are soaked in buttermilk overnight.
- A solution of 5 kg fresh Bilb powder, 1 kg jaggery and 20 litres water is kept for one month followed by applying in root zone to increase the potassium content.
- Pusp Rasayan: Mixture of flowers and cow urine in 1:1 ratio kept for 7 days. Application after 7 days can increase boron content.
- Solution of 2 kg Harada in 10 litres of water followed by *fitkiri* (20 g) is kept for 6 days. Application after 6 days can increase the iron content in soil.
- Application of 2 kg Gaajar Ghaas Svaras in 20 litres of water followed by adding *Fitkiri* powder (20 g) to be kept for 15 days. This application can increase nitrogen content in soil.

REFERENCES

Anbu, V. I., Asokhan, M., Chinnadurai, M., Arunachalam, R., & Balarubini, M. (2018). Potential of indigenous technical knowledge (ITK) through agricultural extension in selected districts of Tamil Nadu, India. *Asian Journal of Agricultural Extension, Economics & Sociology*, 27(2): 1-7.

Pandey, V., Mittal, R., & Sharma, P. (2017). Documentation and application of indigenous traditional knowledge (ITK) for sustainable agricultural development. *Asian Journal of Agricultural Extension, Economics & Sociology*, 15(3): 1-9.

Shailaja, B., Patnaik, H.P. and Mukherjee, S.K. (2012) Assessment of botanicals fermented in cow urine alone and along with panchagavya against brinjal shoot and fruit borer. *Journal of Eco-Friendly Agriculture*, **7**(1): 24-28.

Shreth, C. N., Singh, K. I., William, S. J. (2009) Laboratory evaluation of certain cow urine extract of indigenous plants against mustard aphid, *Lipaphis erysimi* (Kaltenbach) infesting cabbage. *Hexapoda*. 16(1): 11-13.

20 CHAPTER Raising of Vegetable Crops Organically Through Nutrient, Diseases and Pest Management

20.1 INTRODUCTION

Considering a healthy environment as well as for the betterment of our future generation, shifting of conventional agriculture system to organic or sustainable agriculture system is essential. But a farmer or grower is always confused to shift to organic system when it comes to pest and nutrient management. As the chemicals are widely available and act fast as compared to organic agriculture, the farmers are not aware of nutrient and pest management organically. In this chapter, pest and nutrient management in an organic way is described briefly.

20.2 PEST AND DISEASE CONTROL ORGANICALLY

A healthy plant is resistant to many types of disease and pest attack. So, the major aim should be keeping the crop healthy for long term. The strategies to get rid of disease and pest attack are given below:

1. **Healthy planting material:** Planting materials having disease- and weed-free propagules and a certified source is recommended (Latifian 2017).
2. **Resistant varieties:** Resistant varieties which have already adapted to the local environment can prevent the plant from several pests.
3. **Cropping system:** Suitable cropping system like crop rotation, mixed cropping or green manuring plays a major role in organic crop production. Crop rotation prevents the soil-borne pathogen to attack due to lack of host plant. Green manuring can help in increasing the biological activity in soil which increases the beneficial organism population.
4. **Water management:** Proper drainage and irrigation facility should be there which prevents many weed and pathogens to grow in the field. Waterlogging can lead to severe disease infection. Moisture in any plant parts for long time should be avoided.

5. **Soil solarization:** Expose the soil against the hot sunny rays to get rid of harmful microorganisms, weeds and insects is a major practice in organic agriculture. It is normally done in hottest part of the year for a period of four to six weeks. A plastic sheet is covered to increase the temperature of soil.
6. **Field sanitary measures:** Plant residues, stubble, left-over plant materials and infected plant parts should be thrown far away from the field. If not, it may facilitate the pathogens to grow and cause disease.
7. **Animal grazing:** This is an effective method to prevent weeds. Grazing of animals before planting of a new crop makes the field free of various weeds.
8. **Planting time and space:** Proper plant population limits the pathogen to spread from one plant to another plant. It helps in proper aeration which prevents moisture to stay for long term. The time of planting also plays a crucial role as the life stage of pest if coincides with critical stage of the plant then severe disease infection occurs.
9. **Natural enemies:** Conservation of natural enemies is one of the crucial step. Proper habitat should be provided to grow predators (Matyjaszczyk 2015).
10. **Trap crops:** Some crops other than the main crop can be used as trap crop for diverting or trapping of insects, nematodes and pathogens from the main crop. For example, Marigold (*Tagetes* spp.) can be used as a trap crop for nematodes as it traps the nematodes by secreting root exudates.
11. **Roguing:** Physical removal of diseased plants or any other unwanted plants is known as roguing. It is normally profitable in high value crops as it is labour intensive.
12. **Regular monitoring:** Field inspection and regular monitoring is an effective practice for pest prevention through which harmful insects, weeds or diseases symptoms can be identified in early stage. For easy monitoring, light traps, yellow sticky traps, bait traps, water traps, etc., can be used (Hill 2008).
13. **Mechanical pest control:** This method can be applied for insects and weeds which are visible easily. Hand picking of insects and hand weeding through soil solarization, mowing, flaming, tilling or washing is one of the effective alternative of chemical control.
14. **Biocontrol agents:** Using of several potential biocontrol agents is a path towards sustainable agriculture. Biocontrol agents such as *Tricoderma* spp., *Psudomonas* spp, *Agrobacterium* spp., etc.,

can be used to get rid of a wide range of pathogens as well as nematodes. Nuclear polyhedral virus (NPV), natural enemies, predators, etc., can control many major insects in field (Olson, 2008).

15. **Plant protection products in organic farming:** Properly authorized plant protection products of various category such as botanicals, oils, soaps, inorganics can be used for controlling pests. These products can be used directly or diluent. The list of products is given in Table 20.1.

Table 20.1 Plant protection products with their specifications (El-Shafie 2019)

S. No.	Product	Specification
1.	Plant oils	Small bodied insects such as thrips, aphids, whitefly, mites, etc.
2.	Pheromones	Trap and dispenser
3.	Lime	Fungicide
4.	Pyrethrins from chrysanthemum leaves	Insecticides
5.	Sulphur	Fungicide and acaricides
6.	Sodium hypochlorite	Viricide and bactericide
7.	Ethylene	Fruit flies and small bodied insects
8.	Quartz sand	Vertebrate pest
9.	Copper compounds	Fungicide and bactericide (maximum 6kg/annum/ha)
10.	Laminarin	Fungicide and bactericide

20.3 ORGANIC NUTRIENT MANAGEMENT

Soil nutrient can be managed organically by using vermicompost, biofertilizer, green manure, animal manure, microbial and mineral fertilizers. (For production procedure of vermicompost, biofertilizer, green manure refer to practical 2).

1. **Vermicomposting:** The application of vermicomposting depends on the crop. For fruit crops, it is applied near tree basin. For potted plant, it is applied in pot mixture. For nursery or field crops it is applied with soil. For field crops, 5-6 t /ha and 3-5 kg/ plant is applied.
2. **Application of green manure:** If the green manure crop is grown in intercropping system, then harvesting of green manure crop must be done before the sowing of main crop. Green manure crops are if hardy and tall, go for chopping into small pieces and

then incorporate in soil for easy decomposition. Incorporation of green manure crops or harvesting of green manure crops should be done before 2 to 3 weeks of sowing of the next crop to prevent nutrient loss. These crops should not be ploughed deep in soil and is better to apply in surface level only.

3. **Animal manure:** Farm yard manure is one of the best practices to enrich the soil with nutrients. Mixture of dung and cow urine makes a well-balanced nutrient solution. The amount of phosphorus available in soil is equal to the amount present in chemical fertilizers. Nutrient content can be stored properly by focusing on the humidity content of FYM. It should not be more hot or more wet.
4. **Crop rotation:** Rotation of crops like N-fixing crops, or leguminous crops or green manuring crops enrich the nutrient in soil organically in the same land. A proper sequence of crops such as heavy feeder, heavy giver and light feeder should be maintained.
5. **Cover cropping:** These are very important to prevent soil erosion as well as nutrient management. Non-leguminous crops like weeds or grasses recover the mineralized N from soil. Incorporation of these cover crops decomposes in soil with the help of microorganisms and enrich it with several minerals and nutrients.
6. **Application of crop residue:** The nutrient enrichment of soil by crop residue mainly depends on the type of crop residual. For example, cereal straw can enrich 35 kg N/ha while some vegetable crops can add up to 150 kg N/ha.
7. **Biofertilizer:** The microbial biofertilizers play an important role in enriching of soil nutrient. They majorly fixes up N, P and K in soil and makes them available to the plant. It is advised to sow the biofertilizer applied seeds in the evening so that the microorganisms are not be exposed to hot sun radiation.
8. **Organic soil amendments:** Soil amendments such as compost, sphagnum peat moss, alfalfa pellets, wood ash, leaf mould, shredded tree bark can be used to increase soil nutrient. Some other certified soil amendments approved by authorized bodies can also be used for organic nutrient management.
9. **Mineral fertilizer:** These are based on ground natural rock and can only be used as supplement. Sometimes these fertilizers destroy the natural habitats. Normally, lime is used for acidic soils while for alkali soils gypsum is used. Some mineral fertilizers used in organic agriculture are listed in Table 20.2.

Table 20.2 Mineral fertilizers and methods of application (Scialabba 2015)

S. No.	Name of fertilizer	Origin	Methods of application
1.	Lime	Ground limestone	When soil pH is low. Avoid more applications.
2.	Rock phosphate	Pulverized rock	Can mix in compost. Avoid using in red soils
3.	Plant ashes	Burnt organic material	Can mix in compost or around the plant base
4.	Stone powder	Pulverized rock	Mix in FYM

REFERENCES

El-Shafie, H.A.F. 2019. Insect Pest Management in Organic Farming System. In: *Multifunctionality and Impacts of Organic and Conventional Agriculture,* Intech open 1:21.

Hill DS. Pests of Crops in Warmer Climates and their Control. The Netherlands: Springer Science Business Media, B. V; 2008, p. 704.

Latifian M, Rad B. (2017). Efficacy of cultural control for date palm borer management. *Indian Journal of Plant Protection,* 45(1): 7-11.

Matyjaszczyk E. (2015). Prevention methods for pest control and their use in Poland. *Pest Management Science,* 71: 485-491.

Olson S. (2015). An analysis of the biopesticides market now and where it is going. *Outlooks on Pest Management,* 26(5): 203-206.

Scialabba, N. (2015). Nutrient Management in Organic Agriculture. In: *Training Manual for Organic Agriculture.* Food and Agriculture Organization by United Nations. 50-60.

CHAPTER 21 Macro Quality Analysis, Grading, Packaging, and Post-harvest Management

21.1 INTRODUCTION

Organic farming is a way forward to sustainable agriculture system without any harm to the environment. Organic farming not only means to grow the crop in an organic way but also processing and transporting it in the same manner. Different regulatory and certified bodies specify the norms for their processing, packaging and marketing. It is difficult to maintain all the standards such as processing without any chemical additives and more, but consumers health is utmost important above anything.

21.2 MACRO QUALITY ANALYSIS

The definition of quality analysis of organic products keeps changing every time. Prescribed body and certified agencies check the quality and standards of the food and certify it as organic (Hemmerling et al., 2015). Some of the certified bodies are illustrated in Table 21.1.

Table 21.1 Regulatory bodies and their functions

S. No.	Regulatory body	Function
1.	EMS-ISO 14000	Sets guidelines for sustainable production and processing techniques.
2.	HACCP (Codex)- Hazard analysis critical control points	Manages food safety and prevents any risk in processing.
3.	ISO 9000	Quality management system.
4.	SQF 2000CM – Safety quality food	Manage food safety and quality according to the compatibility of ISO 9000 and HACCP Codex.

21.3 GRADING, PACKAGING AND POST-HARVEST MANAGEMENT

21.3.1 Grading and Storing

Quality of the organic products cannot be improved after harvesting is done. Thus, more focus should be given till the product reaches its proper size and stage.

- Harvesting should be done during the coolest part of the day and keep the products in shaded area.
- Handle the products gently.
- Keep the organic products separately. Don't mix with non-organic products.
- Organic products are stored in the certified unit storage.
- Cleaning and disinfection of these products should follow the prescribed regulations. No chemicals should be used for disinfection process.
- Organic and non-organic products cannot be stored and transported together.
- To control all diseases and pests, focus mainly on physical, mechanical and biological methods.
- Chemical use, irradiation, disinfectants are prohibited.
- The transport containers should be washed properly with materials permitted in organic production.

21.3.2 Processing and Post-harvest Management

- For processing, only water, salt may be used. Chemical preservatives and additives are not allowed.
- Some microorganisms and enzymes can be used to some extent but genetically engineered products should be avoided.
- Extraction of organic products can only be done by using water, ethanol, animal and plant oil, vinegar, carbon dioxide, etc.
- Filtration units made up of asbestos should be avoided.
- At the time of processing, more focus should be given on temperature, moisture, chilling injury, sanitation, mixed loads, etc.

21.4 IMPORTANT FACTORS IN POST-HARVEST MANAGEMENT

1. **Temperature**

Table 21.2 Cooling methods to maintain proper temperature (Suslow 2000)

Pre-cooling	Room cooling	Forced air cooling	Hydro cooling	Vacuum cooling	Liquid icing	Chilling injury
1st step	Less efficient	Depends on the rate of air flow and temp	More efficient as removes heat 5 times faster than air	Known as hydrovac cooling as produce is sprayed with water	Mostly effective for dense products which are difficult to cool	For some vegetables
Important for mush-rooms, green beans, sweetcorn, asparagus, brocolli	Proper air circulation around the storage containers	More efficient due to 75-90% fast cooling	Water loss can be reduced	Mainly good for leafy vegetables.	Good for broccoli sweetcorn, etc., having high respiration rate	Very sensitive for basil, cucumber, eggplant, pumpkin, okra, etc.
			Chlorinate can be done to reduce microorga-nisms			

2. Moisture

- To measure the moisture hygrometer can be used.
- Relative humidity of 80-90% is the best range.
- Moisture in the produce also depends on cooling temperature.

3. Sanitation

- Several pathogens such as *Salmonella, Cyclospora, Cryptosporidium, E. coli* can be found in many fruits and vegetables. But use of any type of chemical is banned or permitted to some extent (Table 21.3).
- Chlorine can be used.
- Ethylene produced fruits should kept separately. Otherwise, it may affect some other fruits or vegetables.

Table 21.3 Use of permitted chemicals in post-harvest of organic products (Suslow 2000)

S. No.	Chemical	Function
1.	Acetic acid	Sanitizer or cleanser
2.	Isopropyl alcohol	Used as disinfectant
3.	Hydrogen peroxide	As water or surface disinfectant
4.	Ozone	Produce and equipment disinfection
5.	Bleach	Disinfectant (limited dose)

Contd...

6.	Ammonium sanitizers	Used on non-food contact surface
7.	Peroxyacetic acid	Water, fruit and vegetable surface disinfectant

21.5 POST-HARVEST TREATMENTS (SUSLOW 2000)

- **Wax:** Wood extracted and carnuba derived wax can be used. It should not contain any prohibited synthetic substance. Wax coated products should be well labelled to show use of approved wax.
- **CO_2:** It can be used for crops which can withstand high CO_2 treatment to control against insect pest attack as well as decay loss. It is permitted only for post-harvest management and in controlled atmosphere storage.
- **Fumigants:** Fumigants from the natural source can only be permitted. No chemical fumigations are allowed. Heat vaporized acetic acid is one of the examples.

21.6 PACKAGING

- Materials should be chosen for packaging in such a way that it does not affect the organoleptic character of a product.
- At the time of labelling, don't label water and salt as organic ingredients.
- All raw materials should be written properly on the package along with the percentage.
- All raw materials should be properly labelled according to their source such as organically certified origin or not.
- It should not be labelled as GM (genetically modified) or GE (genetically engineered).

REFERENCES

Hemmerling, S., Hamm, U., Spiller, A. (2015). Consumption behavior regarding organic food from a marketing perspective- a literature review. *Organic Agriculture,* 5: 277-313.

Suslow, T. (2000). Post-harvest Handling of Organic Crops. Organic Vegetable Production in California. 1-10.

22 CHAPTER

Cost of Organic Production System

22.1 INTRODUCTION

Cost of organic production system refers to the sum of all costs-starting from land preparation to marketing. Although the initial cost of production in organic system is high as compared to conventional method. But the profit and return are very impressive in the following years. The interest of consumers in organic products is increasing its demand day by day resulting in more profit and return to the farmers. It was observed that yield can be increased to 25% more compared to conventional farming system. Here, this chapter explains how the organic vegetables return more profit than conventional farming. For reference, only one crop per acre, i.e., maize is taken. Similarly, all products can be calculated in the same way.

22.2 CONCEPT OF COST OF CULTIVATION

To calculate the cost of cultivation, different items and their values are considered and are given in Table 22.1.

Table 22.1 Cost of cultivation in organic system

Cost A_1	• Value of seeds or other planting materials • Value of manure and fertilizers • Value of pesticides • Irrigation cost • Labour cost • Bullock labour cost • Machinery cost • Depreciation on implements • Interest on working capital • Land revenue or other taxes • Miscellaneous expenses
Cost A_2	Cost A_1 + Rent paid on leased land
Cost B_1	Cost A_1 + Interest on own capital assets (excluding land)
Cost B_2	Cost A_2 + Rental value of own land (net land revenue)
Cost C_1	Cost B_1 + Imput value of family labour
Cost C_2	Cost B_2 + Imput value of family labour

22.3 CONCEPT OF INCOME

Farm business income = Gross income - Cost A_2

Family labor income = Gross income - Cost B_2

Net income = Gross income - Cost C_2

Gross income = Value of output

Farm investment income = Net income + Rental value of own land + Interest on owned fixed capital

Calculation

Table 22.2 Value of items involved in organic and conventional farming system

S. No.	Items	Cost of items in organic system (in rupees, per acre)	Cost of items in conventional system (in rupees, per acre)
1.	Human labour		
2.	Bullock labour		
3.	Machine labour		
4.	Seed		
5.	Organic fertilizer		
6.	Organic pesticides		
7.	Miscellaneous		
8.	Interest on working capital		
9.	Interest on fixed capital		
10.	Depreciation		
11.	Rent paid/ lease		
12.	Imputed value of family labour		
13.	Total		

22.4 OBSERVATION

Calculate the cost of cultivation (Table 22.2) and gross return given according to the formula. Prepare a pie chart as shown in Fig. 22.1 by calculating the percentage of items involved in organic farming system. Analyze the profit and cost involved in both methods and compare.

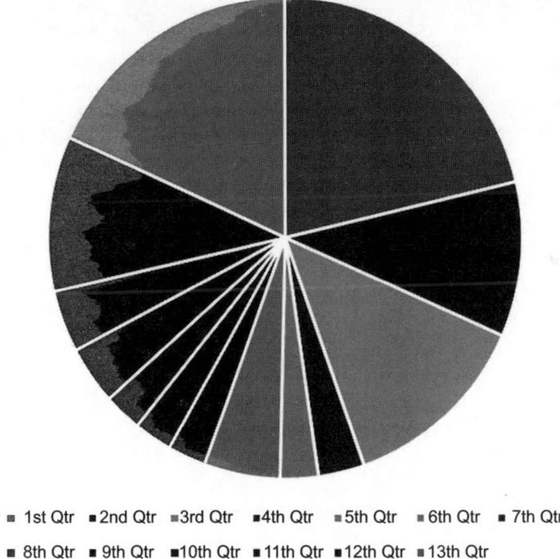

Fig. 22.1 Pie chart showing percentage of items involved in organic farming system (The percentage of items can change considering the location, area, crop, etc.).

22.5 CONCLUSION

The return or profit will be higher in organic system as compared to conventional agriculture. The cost of chemicals used in conventional system will be always higher as compared to organic system. Because in organic system, the substances for plant protection are mostly available in local area or prepared at home.

REFERENCE

Sudheer, P.S.K. 2013. Economics of organic versus chemical farming for three crops in Andhra Pradesh, India. *Journal of Organic Systems*, 8(2): 36-49.

23 CHAPTER Role of Nutrients in the Management of Crop Diseases Through Biopesticides

Shyam Narain Pandey*

Department of Botany, University of Lucknow, Lucknow - 226007, India

* Corresponding author, E-mail: snpandey511@gmail.com

ABSTRACT

The management strategies for high crop production through nutrients balancing and pathogenic disease control are required for sustainable agriculture. The agricultural practices and nutrients management strategies have great potential to minimize the pathogenic disease incidence to the plants. The organic farming practices can fulfil the nutrients requirement of plants through their availability into the soil, and also break host-parasite relationship through altering soil environment. The normal range of nutrients content in soil-plants plays a significant role to increase plant resistance and internal defense system against pathogenic diseases. Use of biocontrol agents for the protection of bacterial and fungal diseases on plants is harmless and eco-friendly. Some microorganisms produce chemicals in rhizosphere facilitate nutrients availability to plant roots for absorption. An extensive assessment of environmental conditions, host-pathogen relations and nutritional status of soil-plant are basic requirement to make disease management strategies. In this chapter, the agricultural practices to change the pathogenic environment and role of essential nutrients in increasing plants resistance against pathogenic diseases and related management strategies have been described.

23.1 INTRODUCTION

The pathogenic disease management through nutrients management and organic agricultural practices may avoid the use of a huge amount of chemical fertilizers and pesticides; it may become a suitable tool for sustainable agriculture. The sustainable agricultural practices are required to arable land of the world in present scenario; these practices are without use of chemical fertilizers and pesticides, and adopting biologically controlled agricultural management system (Marschner 1995; Pandey et al., 2018). The strategies are required to ensure food security for increasing population, human health and socio-economic development for all countries of the world. The sustainable agricultural system is most required modern scenario of researchers to obtain these objectives. The use of chemical fertilizers and pesticides has been increased worldwide for food grain production for achieving food security (Longanathan et al., 2008). But extensive use of chemical fertilizers weakens the soil fertility and decreases quantity and quality of crop produce year after year (Upadhyay and Pandey 2011). In addition, nutritional imbalance and various soil related diseases in plants cause multitude reduction of crop products. The applications of organic manures made from various sources (cow dung, farmyard, poultry, crop residues, etc.) have a great potential to improve soil fertility through favourable soil conditions and nutrients availability (Brady 2001). The improvement in organic manures applied soil fertility is mostly due to improved soil physical conditions (soil texture and structure), water holding capacity, efficient availability of plant nutrients and increased activity of beneficial organisms (Pandey et al., 2018). Thus, the decomposed organic matter increases crop productivity due to improved soil fertility status (Moola et al., 2014) as well as plant health to protect pathogenic diseases.

In comparison to chemical fertilizers, organic sources for nutrition have relatively low content of nutrient content, but more available to the plant roots for absorption (Pandey 2018). Nutrients (macro and micro) from organic resources, after mineralization, remain in the soil for a longer time (Marschner 1986; Brady 2001). Biologically controlled soil management not only provides nutrients regularly (Giri and Pandey 2016), but also enhances growth of microbial population and enzyme activity in the soil for various chemical and biological reactions (Moola et al., 2014). Organic cultivation circulates material in agro ecosystem (Vizzani et al., 2018) and increases crop production with mineral environmental load for keeping ecological balance (Adhikary et al., 2012). The solution to overcome the negative effects of imbalanced use of chemical fertilizers and pesticides in soil-plants health lies in composted organic manures. The essential elements and organic substances (chelates) are absorbed by plant roots in a facilitated manner (Brady and Weil 2002). The nutrients balancing in soil and plants through management strategies, in most of the cases, naturally enhance growth and defense system (Bender et al., 2016; Pandey 2018). Identification of soil conditions and pathogens for specific disease require a large analytical and assessment work to make the management strategies (Oka 2010; Farooq et al., 2011). The diagnosis of visible symptoms appeared on plants, either through pathogenic infection or nutritional imbalances is a significant tool to identification and direction of management strategies.

23.2 SOIL PRACTICES

The agricultural practices by the farmers unconsciously manipulate the soil ecology. The agricultural practices of addition or depletion of organic matter content, extensive cropping, and use of chemicals directly influence soil health and quantity and quality of crops (Sharma 2006; Pandey 2018). The use of crop, tillage practices and addition of decomposed organic matter increase soil health (Peigne et al., 2018), nutrients availability minimizes soil-borne pathogenic diseases (Abawi and Widmer 2000; Sharma 2006). Chemical compounds released from some organic residues into the soil reduce pathogen development (Bailey and Lazarovits 2003). The dynamics of carbon changes due to agricultural practices which affect soil microbial community (Vizzani et al., 2018).

23.2.1 Crop Residues Management

The crop residues contribute organic carbon and mineral nutrients into the soil after decomposition (Federico and Angelini 1991). The rate of decomposition is affected by various factors such as types and amount of crop, interaction between soil organisms (flora and fauna) and time period of decomposition. The pathogen of the soil can be controlled by residue management practices such as crop rotation and tillage lowers the density and survival of pathogen (Edwards 2004). The microbial populations are higher in loamy soil rich in organic matter content in comparison to coarse textured sandy soil. The rate of decomposition of plant residues is grater when buried in soil than when left on the surface of soil. A competition exists between plant roots and soil microorganisms in rhizospheric area for absorption of beneficial and essential elements (Brady and Weil 2002; Pandey et al., 2018). Microorganisms utilize more nitrogen in the form of ammonia in organic matter- rich soil than the plant roots (Brady 2001; Vizzani et al., 2018). The types of crop residues influence the rate of decomposition, and crop rotation varies the quantity and quality of crop residues over the soil surface, consequently, the variations in quantity and quality of decomposed organic matter affects microbial population (Vizzani et al., 2018), pathogen attack condition, nutrients availability and soil fertility (Holm-Nielsen et al,. 2009). Some crop residues release chemical compounds which suppress the growth of pathogens in the soil. Often, plant residues are burnt to limit pest problems in agriculture fields, it also add nutrients to soil but lowers their availability to plant roots by reducing beneficial microorganisms population and organic matter content (Bailey and Lazarovits 2003). The extracts from allelopathic plants contain chemicals, which prevent plants from insect pests (Farooq et al., 2011).

23.2.2 Organic Amendments

A huge amount of agriculture recyclable by-products is generated throughout the world. The bio-products crop residues, cattle manure, poultry manure and municipal solid wastes can be recycled for agricultural use with great potential benefits. The managed recycled organic waste material (Peigne et al., 2018) is not only rich source of nutrients but it also helps to control a large number of pathogens. The meat and bone meal, soy meal and poultry manure at the suitable rate of applications in the soil reduce *Verticillium* wilt, plant parasitic nematodes and common scab of potato. But in most of the cases they also increase the population of bacteria and fungi, some may be beneficial or parasitic in nature. Amendment of compost from major crop residues increases disease incidence on plant roots in soil (Bananomi et al., 2007). Organic manures from plant and animal sources control soil-borne pathogen *Phytophthora cinnamomi* (Aryantha et al., 2000). The production of ammonia and / or nitric acid during the decomposition of organic residues, suppress soil-borne pathogens. The organic matter content changes the pH, nitrification rate and buffering capacity of soil, these changes in soil properties prevent population of parasites in soil (Lazarovits et al., 2001).

23.3 DISEASE MANAGEMENT STRATEGIES

Most of the pathogenic strategies focus on interaction of host-pathogen and environmental conditions. All types of pathogens require suitable conditions like temperature, moisture, humidity for their survival and infection. At any level of pathogenic activity to complete pathogens' life cycle required specific favourable interactions, the breakdown of any step of pathogenic life cycle diminishes their infection in plants. In specific climatic conditions, the majority of strategies are adapted to control the pathogenic infections (Andreote et al., 2017). As soil is a plant growth medium, sometimes becomes unfit for plant growth due to dense population of soil-borne pathogens damage crop and production of compost in the amendment soil from some organic resources, increases suppressiveness of soil-borne pathogens (Janvier and Steinberg 2007). Although, amendment of compost suppress plant pathogens in soil (Hoitink et al., 1997), but increase in pathogenic disease incidence has also been reported (Bonanomi et al., 2007). The study of biological characteristics of soil with the development of molecular methods is an efficient tool for identification of microbial behaviour, and suppressive nature of soil to pathogens (Mazzola 2004). Genotype study modulates soil microbial communities and is helpful in disease control strategies (Bender et al., 2016). Species recognition on the basis of phylogenetic relationship with community is an important tool for prevention of disease (Opik and Davison 2016). Engineering of soil microbiome promotes crop yield as well as helps in disease control strategies (Raaijmakers and Mazzola 2016). Natural mutation, resistant protoclones, genetic engineering and plant tissue culture are some important techniques to develop plant resistance against various pathogens.

23.3.1 Development of Disease-Resistant Varieties

An important alternative to hazardous pesticides use is to develop resistant varieties of crops against pest with variability in the host genetic constitution (Benson et al., 2007). The resistant plants can restrict particular pathogenic disease in a crop. The resistant plants with this altered composition remain disease free through growth habit, toughness of plant parts (stem and leaves), time of maturity and nutrients content. The strategy is very specific and tackles only one or two diseases at a particular time owing to its resistance. Grafting technique of vegetables for developing resistance varieties against soil-borne diseases is a significant process for plant protection (Singh et al., 2017). Some important techniques such as identification of natural mutation, resistant protoclones, genetic engineering and plant tissue culture are used to develop plants resistance against pathogens. The disease resistance conferred by genes like NBS-luciene-rich repeat domains. The miRNAs play role in developing plant resistance against pathogens has been reported (Hussain and Shi 2014). Genome editing technique has been used to develop disease resistance in rice against bacterial blight, and modifying wheat genome for resistance against powdery mildew (Baltes et al., 2017).

23.3.2 Cultural Control

Preventive strategies are more suitable in organic farming system than curative. It is agricultural practices for sufficient nutrients supply to promote healthy plants. The cultural habits like weeds management, sanitation and right planting time reduce the density and favourable conditions for growth of pathogens. Crop rotation practices also minimize the disease incidence (Van Bruggen and Finckh 2016). The selection of crop species and variety, the temporal and spatial pattern of crop rotations are practices which minimize the interaction between susceptible host and virulent pathogen (Stockdale et al., 2000). Soil solarisation of agricultural fields reduces soil-borne inoculations.

23.3.3 Application of Organic Amendment

The recyclable by-products from biological resources are used in the agricultural field. Their use with great potential benefits, provides nutrients and minimizes soil-borne pathogens. The amendment of vermicompost, soy meal and poultry manure improve nutritional as well as soil conditions (Vizzani et al., 2018) which support healthy plant growth and defense against parasites (Aryantha et al., 2000). When partially decomposed organic resources applied in the soil invite fungal and bacterial pathogens, cause diseases and damage crop growth and yield. (Bonanomi et al., 2007). The decomposition process produces some chemicals (nitrous acid and ammonia) which suppress fungal pathogens has been reported (Lazarovits et al. 2001).

23.3.4 Use of Agro-Products

Plant extracts, essential oils, fermented buttermilk and cow urine are some of the controlling measures of pests and diseases in organic farming system, in India. A large number of plant extracts are used to reduce incidence of disease. Products such as extract and essential oils obtained from eucalyptus, cinnamon, clove, *Azadirecta indica*, garlic, basil and oleander, etc., control a large number of fungal diseases. The use of such plant products is eco-friendly and non-toxic to food web (Cosic et al., 2010; Nashwa and Abo-Elyousr, 2012). The essential oils and plant extracts are anti-pathogenic in nature and control various soil-borne diseases (Nashwa 2017).

23.3.5 Application of Biocontrol Agents

Bioagents are relatively host specific and safe in disease management because they do not disturb biotic systems (Srinivas and Ramakrishna 2005). Biocontrol agents such as *Pseudomonas* sp., *Bacillus* sp. and *Trichoderma* sp. are compatible with most other pest control methods. They are harmless and control various pathogens by their action competition, antibiosis, antagonism, increased nutrient content uptake and stimulation of host resistance (Kloepper et al., 1997). Commercial bioagents can be locally produced and used in agricultural areas. The bioagents not only kill the specific pathogens but also, sometimes, promote the growth of plants. A biocontrol agent *Streptomyces* suppresses damping-off disease in tomato (Sabaratnam and Traquair 2002). Actinomycetes in soil, control leaf and shoot blight in *Eucalyptus* (Himaman et al., 2016). Some other biocontrol agents suppress plant diseases are *Bacillus* sp., and *Metarhizium anisopliae* against many insect pests has been reported (Zimmermann 1993). To control many fungal and bacterial plant diseases, herbal pesticides and use of biocontrol agents are very effective both for soil and air-borne diseases (Giri and Pandey 2016).

23.4 NUTRIENTS MANAGEMENT STRATEGIES

The goal of soil management is to must keep nutrient cycles in a balanced way. Nutrients management maintains the soil's capacity to regular supply of nutritional needs of plants and ultimately to all animals and human beings. The management of soil must be focused on the causes of nutrients loss such as leaching and run-off (system leakage), monoculture, extensive cropping, etc. The agriculturists aim nutrientmanagement (Sharma 2006; Pandey 2018) to increase crop production, quantitative and qualitative, and plant protection from various pathogens (Halsall et al., 1983) to increase their incomes. The soil nutrients conservation is a significant process of management. It includes renewal or reuse of the resources and nutrient budgeting (balance between system inputs and outputs). In the recycling process nutrients are returned to the same land from which they were removed. Recycling from litter fall and leaving crops residue in the field are example of nutrient management (Vizzani et al., 2018). The management of nutrition in the soil needs integrated management of physical, chemical and biological processes. The components of integrated management are tillage on organic matter accumulation (Peigne et al., 2018), nutrients availability, and use of bioagents for mineralization and protection of fire on vegetation land are significant. Nutrients management also affects environment, particularly water quality. Nutrient management plans mostly focus on nitrogen and phosphorus, but other macro- and micronutrients require their management for optimum plant production (Sharma 2006). The nutrient management plan is a set of guidelines for application of nutrient sources to land. The some important accessories of this plan are preparation soil map, crop rotation; soil tests guided nutrients application, plant tissue analysis results, nutrient analysis of manure or other soil amendments, realistic yield goals and a complete nutrient budget for nutrients elements in the plant-production systems (Brady and Weil 2002).

23.5 MINERAL NUTRITION AND DISEASE PROTECTION STRATEGIES

23.5.1 Role of Macronutrients

The role of mineral nutrition in growth and metabolism of plants are well explained (Sharma 2006; Pandey 2018). Essential mineral nutrients (micro and macro) also affect the resistance of plants through various plant breeding and biotechnological approaches, a large number of plants evolved in resistance against diseases and pests. Resistance in a plant increases through thicker epidermal cells and high deposition of lignin. Plants also develop protection measures through biochemical changes under the influence of nutrients contents, secrete inhibitory or repelling substances. Plant responses to pathogens attack by increased level of lignifications and phytotoxins (Nobel and Coventry 2005).

Resistance in a plant is controlled by specific genes and also influenced by various environmental factors. Manipulation of plants resistance power through essential nutrients as environmental factor is relatively simple as compared to applications of pesticides and fungicides (Perez et al., 2011). Plants with optimum nutritional concentration have the highest resistance to pathogenic attack (Marschner 1986). The phenolic compounds and flavanoids content depend on mineral nutrients regulated metabolisms (Sharma 2006). Most of the parasitic pathogens secrete pectolytic enzymes to penetrate the epidermal cells. These enzymes dissolve the middle lamella strongly suppressed by the Ca^{2+}which protect the plants from pathogens (Marschner 1995). Some fungal diseases suppressed by inorganic salts have been reported (Deliopoulos et al., 2010).

23.5.1.1 Role of Nitrogen and Potassium in Disease Control

The role of nitrogen and potassium in modulating disease resistance has been demonstrated (Marschner 1995). But their doses are controversial for effect on obligate and facultative parasite. The high content of nitrogen in soil promotes infection of facultative parasite (Alternaria, Fusarium) and also some bacterial infections in plants (R). The high concentration of potassium increases the resistance power of plants against both obligate and facultative parasites. The obligate parasites cause rust disease in susceptible wheat plants, its severity increases with high nitrogen supply in growth medium, while the plants deficient in nitrogen are most resistant. In tomato, bacterial leaf spot disease caused by a facultative parasite shows suppressed disease at elevated nitrogen supply. The decrease in potassium content in soybean pods, increases pod blight disease. Supply of potassium supresses this disease (Marschner 1995). The elevated nitrogen content enhances the severity of dieback disease (Scarlett et al., 2013).

23.5.1.2 Role of Calcium and Other Mineral Nutrients

The incidence of pathogens decreased by calcium content in plant tissues has been reported (Muchovej and Muchovej 1982). Many bacterial and fungal parasites produce proteolitic enzymes (such as Polygalactoronase) to dissolve middle lamella is inhibited by Ca^{2+} content in tissues (Serrano et al., 2012). The sufficient calcium in plant tissue protect the 'twin stem' abnormality in soybean (necrosis of apical meristem occurs). This abnormality is endemic in acid tropical soils, at the same time soybean infected with Sclerotium species is also protected by calcium supply (Muchovej and Muchovej 1982). The calcium concentration in xylem sap protects the walls of vessels to various parasitic fungi leads to plugging of the vessel and causes wilting symptoms (Fusarium wilt). Before storage of fleshy fruits, the treatment of calcium is an effective procedure for preventing physiological disorders and from fruit rotting caused by fungal pathogens (Marschner 1986). The application of chloride and sulphate salt of calcium controls pistachio gummosis casual agent *Phytophthora pistaciae* (Mostowfizadeh- Ghalamfarsa et al., 2018). The poor defence mechanism due to nutrients deficiency is less effective for protection of pathogens (both obligate and facultative parasites) (Sugimoto et al., 2005).

23.5.2 Protective Role of Micronutrients in Plants

The role of micronutrients to protect plants from various physiological disorder through influence the generation and detoxification of highly reactive oxygen species, they are formed due to biotic and biotic stresses (Pandey 2018). The antioxidative enzymes have micronutrient cofactor which participates in antioxidant defense mechanisms (Sharma 2006) against reactive oxygen species under the stress conditions either by pathogens or nutrients imbalances (Pandey et al., 2002). Reactive oxygen species are formed in mitochondria and chloroplasts (Elstner 1991). The enzyme superoxide dismutases have various micronutrient cofactors such as Fe (iron), Mn (manganese), Cu (copper) and Zn (zinc), help at first line of defence against reactive oxygen species (Alscher et al., 2002).

The cytosol and apoplasm, superoxide ions (O^-_2) are converted to hydrogen peroxide (H_2O_2) by the activity of Cu-Zn SOD. The H_2O_2 then is converted in to harmless water molecules by the activity of ascorbate peroxidase and catalase (haem enzyme). Thus, as cofactors of enzymes catalyses, superoxide dismutase and ascorbate peroxidase, some micronutrients such as Mn, Fe, Zn and Cu contribute to their role in defence against oxidative stress (Pandey et al., 2018). In transgenic plants, overexpression of superoxide dismutases increases tolerance to oxidative stress has been reported (Slooten et al., 1995; Perl et al., 1993).

The micronutrients also play their role in protection of plants against pathogenic infections (Graham 1983; Graham and Webb 1991; Huber and Graham 1999). Polymerization of phenolic compounds to lignins and suberine in the presence of copper, boron and manganese, which strengthens the plant cell walls, making plants less susceptible to penetration of pathogens. Micronutrients like Zn, Cu, Mn, B and Fe provide protection to various plants against infections from pathogens through producing H_2O_2 which functions as cytotoxins (Lamb and Dixon 1997). The production of H_2O_2 also induces defence system of the host against pathogenic infections (Orozeo-Cardinas et al., 2001). Copper is constituent of an enzyme diamine oxidases which generates and regulates H_2O_2 concentration in response to stresses.

Iron protects from reactive oxygen species, salt stress and pathogens (Sharma 2006). As a constituent of Fe-SOD, iron detoxifies the O^-_2 by dismutating to H_2O_2. Other iron containing enzymes ascorbate peroxidase and catalase scavenging H_2O_2 in the cell (Sharma 2006). An iron containing enzyme alternative oxidase provides an alternate pathway to the reducing equivalents from quinol and prevents them from interacting with oxygen to generate reactive oxygen species during mitochondrial electron transport chain (Millenear et al., 1996). The heam enzymes are reduced in polymerization of the cinnamyl alcohols to form lignin which contributes to resistance in plants against pathogen attack (Hartified and Vermerris 2001). The peroxidase a heam enzyme, inhibits the germ tube growth of phytopathogenic fungi (Caruso et al., 2001).

Manganese, a co-factor of superoxide dismutase (Mn-SOD) is a significant enzyme in antioxidant defence system (Pandey 2018) in mitochondria (Bowler et al., 1991). Manganese is a micronutrient provides resistance to plants against pathogens (Graham 1983). The role of manganese in crop resistance and to tolerance to various diseases has been reported (Huber and Graham 1999).

In Australia, the 'take-all disease' in wheat is due to low concentration of manganese in plants rendered them susceptible to the disease involvement of manganese in biosynthesis of lignin and phenolics in roots make a plant resistance to take all disease. But, pattern in biosynthesis is variable in genotypes (Rengel et al., 1994). The synthesis of lignin and phenolics under suitable concentration of manganese in plants, inhibits the penetration of the pathogens in to the root cells and protects against parasitic fungi. Manganese protects the pathogenic diseases such 'take all' in wheat (Graham and Rovira, 1984), 'rust' in soybean (Morab et al., 2003) and brown spot of rice (Kaur et al., 1979) have been observed.

Copper plays its role in protection against superoxide ions accumulated near the pigment system I (PS I) complex (Ogawa et al., 1995) through becoming as a constituent of Cu-Zn superoxide dismutase. The plants are also protected by pathogenic diseases with sufficient copper content which lignifies tissues (Graham and Klebb 1991) copper deficient plants are more vulnerable to pathogenic infections. For protection against pathogenic diseases in plants copper plays a role through enhanced protection of reactive oxygen species. The reactive oxygen species are cytotoxic to pathogens, and also induce defence system in host (Lamb and Dixon 1997). Copper diamine oxidases activates defence mechanism in plants (Rea et al., 2002). The wounding and infection by *Ascochyta rabiei* (pathogen) active copper amine oxidase induces enhanced production of H_2O_2

which trigger common visible symptoms such as reduction in leaf size and condensation of shoot growth (Sharma 2006).

Zinc deficiency adversely affects activity of superoxide dismutase (Pandey et al., 2002) particularly Cu-Zn SOD is most sensitive enzymes to zinc deficiency (Yu and Rengel 1999; Pandey 2018). Whereas, low zinc- deficient leaves show increased activity of iron containing SOD (Fe- SOD) (Bonnet et al., 2000). Low zinc level and poor growth of plants lose resistance against pathogenic diseases (Sharma 2006). The role of molybdenum in nitrogen metabolism during nitrogen fixation by free living microorganisms and symbiotic bacteria which promote plant growth have been reported (Sigel and Sigel 2002; Sharma 2006) to promote defence system in plants (Lamb and Dixon 1997). In the response of zinc deficiency, plants show poor growth. Molybdenum is a co-factor of several enzymes in bacteria and plants. At adequate supply of molybdenum help symbiotic bacteria (*Rhizobacterium leguminasorum*) in the root nodules for nitrogen fixation. Molybdenum also promotes nodulation and its development due to infection of microorganisms for nitrogen fixation (Brodrick et al. 1992).

23.6 CONCLUSION AND FUTURE PROSPECTS

Organic farming system is most acceptable in present scenario of sustainable agriculture. To obtain optimum crop production bioresources provide various macro and micro essential nutrients in the soil, they must be available in sufficient quantity for plant's absorption. The protection of plants from various pathogenic diseases is also very important. The amendment of organic residues in fully decomposed form, use of bioresources, bioagents and biological practices can provide solutions for both nutrition and disease protection for plants. Nutrients when present in their normal critical range in soil-plant system are very helpful for plant protection from incidence and development of diseases in various ways. The development of model for balancing nutrients and applications of bioagents and bioresources to achieve uncontaminated environment, agriculture and human health is an urgent need for future prospects.

REFERENCES

Abawi, G.S.; Widmer, T.L. 2000. Management practices on soilborn pathogens, nematodes and root diseases of vegetable crops. *Applied Soil Ecology*, 15: 37-47.

Adhikari, B.N.; Savory, E.A.; Vaillancourt, B.; Childs, K.L.; Hamilton, J.P.; Day, B.; Buell, C.R. 2012. Expression profiling of *Cucumis sativus* in response to infection by *Pseudopernospora cubensis*. *PLoS ONE*, 7, e34954, https//doi.org/10.1371/journal. pone.0034954

Alscher, R.G.; Erturk, N.; Health, L.S. 2002. Role of superoxide dismutases (SOD) in controlling oxidative stress. *J. Exp. BOT*, 53, 1331-1341.

Andreote, F.D.; Michele de Cassia; Pereira e Silva. 2017. Microbial communities associated with plants: learning from nature to apply it in agriculture. *Current Opinion in Microbiology*, 37: 29-34.

Aryantha, I. P.; Cross, R.; Guest, D. I. 2000. Suppression of *Phytophthora cinnamomi* in potting mixes amended with uncomposted and composted animal manures. *Phtopathology*, 90: 775-782.

Bailey, K. L.; Lazerovits, G. 2003. Suppressing soil born diseases with residue management and organic amendments. In: Soil and Tillage Research, 72: 169-180. Doi: 10.1016/S0167- 1987 (03)00086-2

Baltes, N.J.; Humanes, J.G.; Voytas, D. 2017. Genome engineering and agriculture: opportunities and challenges. *Progress in Mol. Biol. Translation Sci*, 149: 1-26. Doi.org/ 10.1016/bs.pmbts.2017.03.011

Bender, S.F.; Wagg, C.; Maecel, G.A.; Heijden, V.D. 2016. An underground revolution: biodiversity and soil ecological engineering for agriculture sustainability. *Trends in Ecology and Evolution*, 31: 440-445.

Benson, D.A.; Karsch-Mizrachi L.; Lipman, D.J.; Ostell, J.; Wheeler, D.L. 2007. Gene Bank, *Nucleic Acids Research*, 35: 21-25. Doi: 10.1093/ nar/gkl986

Bonanomi, G., Antignani, V., Pane, C., Scala, F. 2007. Suppression of soilborne fungal diseases with organic amendments. *J. Plant Pathology*, 89(3): 311-324.

Bonnet, M.; Camares, O.; Veisseire, P. 2000. Effect of zinc and influence of *Acremonium lolii* on growth parameters, chlorophyll a, fluorescence and antioxidant enzyme activities of ryegrass (*Lolium perenne* L.). *J. Exp. Bot*, 51: 945-953.

Bowler, C.; Slootan, L.; Vanderbnanden, S.; De Rycke, R.; Botterman, J.; Sybesma, C.; Van Montagu, M.; Inze, D. 1991. Manganese superoxide dismutase can reduce cellular damage mediated by oxygen radicals in transgenic plants. *EMBOJ*, 10: 1723-1732.

Brady, N.C. The Nature and Properties of Soils. 2001, 10th ed., Printice-Hall of India, New Delhi

Brady, N.C., Weil, R. The Nature and Properties of Soils. 2002, (13th ed.), Pearson Education (Singapore) Pvt. Ltd., Delhi.

Brodick, S. J.; Sakala, M. K.; Giller, K. E. 1992. Molybdenum reserves of seed, and growth and nitrogen fixation by (*Phaseolus vulgaris* L.). *Biol. Fert. Soils*, 13: 39-44.

Caruso, C.; Chilosi, G.; Leonardi, L.; Birtini, I.; Magro, P.; Buonocore, V. 2001. Caporable. A basic peroxidase from wheat kernel with antifungal activity. *Phytochemistry,* 58: 743-750.

Deliopoulos, T.; Kettlewell, P. S.; Hare, M. C. 2010. Fungal disease suppression by inorganic salts: *A review. Crop Protection,* 29: 1059-1075.

Edwards, S.G. 2004. Influence of agricultural practices on *Fusarium* infection of cereals and subsequence contamination of grain by Trichothecene mycotoxins. *Toxicology Letters,* 153: 29- 35.

Elstner, E.F. 1991. Mechanism of oxygen activation in different components of plants In: Active oxygen species, oxidative stress and plant metabolism. *Am. Soc. Plant Physiol.,* Rockville MD, USA, 13-25.

Farooq, M.; Jabran, K.; Cheema, Z. A.; Wahid, A.; Siddique, K.H. 2011. The role of allelopathy in agricultural pest management. *Pest Management Science,* 67: 493-506. http://doi.org/10.1002/ps.2091

Federico, R.; Angelini, R. (1991). Polyamine catabolism in plants. In: *Biochemistry and Physiology of Polyamines in Plants* (R.D., Solcum, H. E., Flores (eds., CRC Press. Boca Raton, FL, 41-56.

Giri, P.; Pandey, S.N. 2016. *Trichoderma* spp. And metal chelator enhanced growth and yield in wheat plants by increasing zinc availability. *International Education and Research Journal,* 2: 66-70.

Graham, R.D. 1983. Effect of nutrient stress on suseptability of plants to disease with particular reference to trace elements. *Adv. Bot. Res,* 10: 221-276.

Graham, R.D.; Rovira, A.D. 1984. A role of manganese in the resistance of wheat plants to take all. *Plant Soil,* 78: 441-477.

Graham, R.D.; Webb, M.J. 1991. Micronutrients and plant disease resistance and tolerance in plants. In: *Micronutrients in Agriculture.* Mortvedt, F. R., Cox, L. M., Suman, Welch, R. M. (eds). Sol. Sci. Soc. Amer. Book Ser. No. 4, Medison, WI., 329-370.

Halsall, D.M.; Forrester, R.I.; Moss, T.E. 1983. Effect of nitrogen, phosphorus and calcium nutrition on growth of eucalypt seedlings and on the expression of disease associated with *Phytophthora cinnamomi* infection. *Aust J. Bot,* 31: 341-355 doi. org/10.1007/&13313-012-0173-2

Hartfied, R.; Vererris, W. 2001. Lignin formation in plants. The dilemma of linkage specificity. *Plant Physiol,* 126: 1351-1357.

Himaman, W.; Thamchaipenot, A.; Pathon-aree, W.; Duangmal. K. 2016. Actinomycetes from *Eucalyptus* and their biological activities for controlling *Eucalyptus* leaf and shoot blight. *Microbiological Research,* 188: 45-52.

Hoitink, H.A. J.; Stone, A.G.; Han, D.Y. 1997. Suppression of plant diseases by composts. *Hort. Science,* 32: 84-187.

Holm- Nielsen; Seadi, T.A.; Oleskowicz- Popiel, P. 2009. The future of anaerobic digestion and biogas utilization, *Bioresource Technology,* 100: 5478- 5484.

Huber, D.M.; Graham, R.D. 1999. The role of nutrition in crop resistanceand tolerance to diseases. In : Mineral Nutrition of Crops. Fundamental Mechanisms and Implications. Rengel Z. (eds). Haworth Press, New York, 169-204.

Hussain, S.S.; Shi, B. 2014. Role of miRNAs in abiotic and biotic stresses in plants. In: Emerging Technologies and Management of Crop Stress Tolerance, 1: 181-207. Doi.org/10.1016/B978-0-12-800876-8.00009-6

Janvier, C.; Steinberg, C. 2007. Soil health through soil disease suppression: which strategy from descriptors to indicators? *Soil Biology and Biochemistry,* 39: 1-23.

Kaur, P.; Kaur, S.; Padmanabhan, S.Y. 1979. Effect of manganese and iron on incidence in brown spot disease of rice. *Indian Phytopath,* 32: 287-288.

Kloepper, J.W.; Tozun, S.; Zehnder, G.W.; Wei, G. 1997. Multiple disease protection by rhizobacteria that induce systemic resistance-historic precedence. *Phytopathology,* 87: 136-137.

Laganathan, P.; Hedly, M.J.; Grace, N.D. 2008. Pasture soils contaminated with fertilizer-derived Whitacre, D.M. cadmium and fluorine: Livestock effects. In Whitacre, D.M. (eds.), Reviews of Environmental contamination and Toxicology, Springer, New York, NY, 192: 29-66.

Lamb, C.; Dixon, R.A. 1997. The oxidative burst in plant disease resistance. Annu. Rev. *Plant Physiol. Plant Mol. Biol.,* 48: 251-275.

Lazarovits, G. Tenuta, M.; Conn, K.L. 2001. Organic amendments as a disease control strategy for soilborne diseases of high value agricultural crops. *Australisian Plant Pathology,* 30: 111-117.

Marschner, H. (1986). Mineral nutrition of higher plants. Academic Press.

Mazzola, M. 2004. Assessment and management of soil microbial community structure for disease suppression. Annu. Rev. *Phytopathol.,* 42: 35-59.

Millenaar, F.F.; Benschop, J.J.; Wagner, A.M.; Lambers, H. 1996. The role of alternate oxidase in stabilizing the *in vitro* reduction state of the ubiquinone pool and the activation state of the alternate oxidase. *Plant Physiol,* 118: 599-607.

Moola, R.; Davari, M.R.; Sharma, S.N. 2014. Direct, residual and cumulative effects of organic manures and biofertilizers on yields, NPK uptake, grain quality and economics of wheat (*Triticum aestivum*). *J. Organic Systems,* 9: 16- 30.

Morab, H.; Koti, R.V.; Chetti, M.B.; Patil, P.V.; Nalini, A.S. 2003. Role of nutrients in inducing rust resistance in soybean. *Indian J. Plant Physiol,* 8: 85-88.

Mostowfizadeh-Ghalamfarsa, R.; Hussaini, K.; Ghasemi- Fasaei. R. 2018. Effects of calcium salts in controlling *Phytophthora pistaciae,* the causal agent of pistachio gummosis, *Eu. J. Plant Pathology,* 151: 475-485.

Muchovej, R.M.C.; Muchovej, J.J. 1982. Calcium suppress of Sclerotium-induced twin stem-Abnormality of soybean. *Soil Science,* 134: 181-184.

Nobel, R.; Coventry, E. 2005. Suppression of soil-born plant diseases with compost. *Biocontrol Science and Technology,* 15: 3-20.

Ogawa, K.; Kanematsu, S.; Takabe, K.; Asada, K. 1995. Attachment of Cu-Zn superoxide dismutase to thylakoid membranes at the site of superoxide generation (PS I) in spinach chloroplasts: Detection by immune-gold labeling after rapid freezing and substitution method. *Plant Cell Physiol,* 36: 565-573.

Oka, Y. 2010. Mechanisms of nematode suppression by organic soil amendments- a review. *Applied Soil Ecology,* 44: 101-115.

Opik, M.; Davison, J. 2016. Uniting species- and community-oriented approaches to understand arbuscular mycorrhizal fungal diversity. *Fungal Ecology,* 24: 106-113.

Orzeo- Cardinas, M.; Narvaez- Vasquez, J.; Ryan, C. 2001. Hydrogen peroxide acts as a second messenger for the induction of defense genes in tomato plants in response to wounding, system in and methyl jasmonate. *Plant Cell,* 13: 179-191.

Pandey, N.; Pathak, G.C.; Singh, A.K.; Sharma, C.P. 2002. Enzymic changes in response to zinc nutrition. *J. Plant Physiol,* 151: 1151-1153.

Pandey, S.N. 2018. Molecular Functions of Micronutrients Towards Abiotic Stress Tolerance in Plants. In: Hasanuzzaman, M., Fujita, M. Oku, H. Nahar, K., Hawrylak-Nowak, B. (eds.) *Plant Nutrients and Abiotic Stress Tolerance*, Springer, Singapore. https://doi.org/10.1007/978-981-10-9044-8_6

Pandey, S.N.; Abid, M.; Abid Ali Khan, M.M. 2018. Diversity, Functions, and Stress Responses of Soil Microorganisms In: Egamberdieva, D., Ahmad, P. (eds.). Plant Microbiome: Stress Response. Microorganisms for sustainability, Vol. 5, Springer, Singapore.

Perez-Garcia, A.; Romero, D.; Vicente, A.D. 2011. Plant protection and growth stimulation by microorganisms: Biotechnological applications of Bacilli in Agriculture. *Current Openion in Biotechnology*, 22: 187-193. Doi.org/10.1016/j.copbio. 2010.12.003

Perl, A.; Perl-Treves, R.; Galili, S.; Aviv, D.; Shalgi, E.; Malkin, S.; Galun, E. 1993. Enhanced oxidative stress defence in transgenic potato expressing tomato Cu, Zn superoxide dismutase. *Theor. App. Gen.*, 85: 568-576.

Raaijmakers, J.; Mazzola, M. 2016. Soil immune responses. *Science*, 352: 1392-1393.

Rea, G.; Laurenzi, M.; Tranquilli, E.; D'Ovidio, R.; Federico, R.; Angelini, R. 1998. Developmentally and wound regulated expression of the gene encoding a cell wall copper amino oxidase in chickpea seedlings. *FEBS Let*,.437: 177-182.

Rengel, Z.; Graham, R.D.; Pedler, J.F. 1994. Time cover of biosynthesis of phenolics and lignin in roots of wheat genotypes differing in manganese efficiency and resistance to take all fungus. *Ann. Bot.*, 74: 471-477.

Sabaratham, S.; Traquair, J.A. 2002. Formulation of Streptomyces biocontrol agent for the suppression of rhizoctonia damping-off in tomato transplants. *Biological Control*, 23: 245-253.

Scarlett, K., Guest, D.; Daniel, R. 2013. Elevated soil nitrogen increases the severity of dieback due to *Phytophthora cinnamomi, Australasian Plant Pathol.*, 42:155–162.

Serrano, M.S.; De Vita, P.; Fernandez-Rebollo, P.; Hernandez, M.E.S. 2012. Calcium fertilizers induce soil suppressivness to *Phytophthora cinnamomi* root rot of auercusilex. *European J. Plant Pathology*, 132: 271-279.

Sigel, A.; Sigel, H. 2002. Molybdenum and Tungsten. Their roles in biological systems, Marcel Dekker, New York.

Singh, H.; Kumar, P.; Chaudhari, S.; Edelstein, M. 2017. Tomato Grafting: A Global Perspective. *Hort Science*, 52:1328-1336. Doi:10.21273/HORTSCI11996-17

Sharma, C.P. 2006. Plant Micronutrients, Science Publishers, NH, USA.

Slooten, L.; Capiau, K.; Van Camp, W.; Van Montagu, M.; Sybesma, C.; Inze, D. 1995. Factors affecting the enhancement of oxidative stress tolerance in transgenic tobacco over expressing manganese superoxide dismutase in the chloroplasts. *Plant Physiol.,* 107: 735-750.

Srinivas, P.; Ramakrishna, G. 2005. Biological management of rice seed born pathogens by native biocontrol agents. *Annals of Plants Prote. Sci.,*13: 422-426.

Stirling, G.R. 2017. Biological control of plant parasitic nematodes. *Disease of Nematodes,* 1: 103-150.

Stockdales, E. 2000. Agronomic and environmental implications of organic farming systems. *Advances in Agronomy,* 70: 261-327.

Sugimoto, T.; Airoo, M.; Sugimoto, M.; Watanabe, K. 2005. Reduction of Phytophthora stem rot disease on soybeans by the application of $CaCl_2$and $Ca(NO_3)_2$. *J. Phytopathology,* 153: 536-543. doi.org/10.111/j.1439-0434.2005.01016

Upadhyay, A.; Pandey, S.N. 2011. Reclamation of sodic soil by organic manure and medicinal plant (*Trachyspermum ammi L.*). *Res. Environ. Life Sci.,* 4(2): 63-68.

Van Bruggen, A.H.C.; Finckh, M.R. 2016. Plant disease and management approaches in organic farming systems. *Annu. Rev. Phytopathol.,* 54: 25-54.

Yu, Q.; Rengel, Z. 1999. Micronutrient deficiency changes activities of superoxide dismutase and ascorbate peroxidase in tobacco plants. *J. Plant Nutrition,* 21: 1427-1437.

Zimmermann, G. 1993. The entomopathogenic fungus Metarhizium anisopliae and its potential as a biocontrol agent. *Pesticide Management Science,* 37: 375-379. Doi.org/ 10.1002/ps.2780370410

24 CHAPTER Zero Budget Natural Farming: Where Healthy Soil Makes Healthy Crop—An Eco-friendly Approach to Manage Insect Pests and Diseases

G. Venkat Raman

ZBNF Wing, Raithu Sadhikara Samstha, Department of Agriculture.

Gunter-522034, Andhra Pradesh, India.

Corresponding author: govenkatraman@gmail.com

ABSTRACT

Zero budget natural farming (ZBNF) is an alternative farming system to present high cost chemical agriculture. It addresses soil degradation, pollution, human health and climate change issues. 365 days green cover which is a new paradigm in natural farming is discussed in this chapter. ZBNF core principles by emphasizing four wheels, viz., Beejamruth, Jeevamruth, Achadan and Waphasa are described scientifically. The role of healthy soils in producing healthy crops is the core subject of this chapter. How natural farming practices improve the soil microbial diversity, humic substances and the role of useful microbes in inducing systemic resistance is mentioned by quoting research evidence. As ZBNF is similar to agroecosystem principles, the cropping systems with poly crops intercrops, trap crops and boarder crops effectively manage pests and diseases. The list of botanical pesticides and cow based solutions is provided along with non-negotiables for plant protection.

24.1 INTRODUCTION

Zero budget natural farming (ZBNF) is a holistic alternative to the present paradigm of high cost chemical inputs based agriculture and to address the negative and uncertain impacts of climate change. ZBNF principles are in sync with the principles of Agroecology, a worldwide movement. This farming synergises with nature and uses the natural inputs of marginal or low cost that can be made by farmer himself using natural substances.

Natural farming is done without using any chemical fertilizers and pesticides. The farming system is based on the natural principles of biodiversity, both above the ground and below the ground. Crop diversity with complete green cover or mulching are important practices besides periodic application of Jeevamruth.

Often people misunderstand both organic and natural farming systems are same, but there is big difference between these two systems. Natural farming does not consume any external inputs and incur any cost on inputs, except jaggery and pulse flour. It is purely based on preparations of desi cow products (urine and dung). Whereas organic farming consumes external on chemical inputs like vermicompost, biofertilizers, FYM and other non-chemical pest control inputs.

In the ZBNF farming systems, plant protection is not an important activity, but soil nutrient management is a major activity. As a holistic approach, cowbased inputs and poly cropping models play an important role in soil nutrient management and plant protection. This chapter explains the role of healthy soil in making healthy crops. How agroecosystems manage pest populations naturally and role of eco-friendly concoctions in minimising pest and disease incidence are described in this chapter.

24.2 IMPORTANCE OF ZBNF

ZBNF eliminates the use of chemical inputs. Eco-friendly farming practices with polycropping, layered plantations, covering the green cover round the year, etc., absorbs the atmospheric carbon and sinks into the soil. It addresses sustainable development goals.

365 days green cover decreases the global temperature and creates atmosphere for timely rains. It provides environment for microbes to sustain in the soil and also restores the entire soil biodiversity.

AP-ZBNF programme is promoting integrated farming systems (IFS) which utilizes resources efficiently and becomes livelihood source for small and marginal farmers. Agro-forestry models like 5-layer models restore the diversity, ensure the food and nutrition security to entire family.

The water vapour present in the air is 10 times more than that of rivers and underground water. Unfortunately, this water is not used for agriculture. The Waphasa in ZBNF is the only solution to absorb such water vapours for agriculture. This is very important to tackle climate change and drought conditions.

24.3 FOUR WHEELS OF ZBNF

ZBNF practices are built on four core principles, as explained below:

24.3.1 Beejamruth

It is the composition of cow urine and cow dung. This has a number of microbial species. Seed coating of this is to protect the seed or seedling from pests and diseases and promote the healthy germination process.

By the treatment of Beejamruth, coating of microorganisms is done around the seed.Farmers are advised to treat seeds before sowing.

24.3.2 Jeevamruth

Jeevamruth is applied to soil as well as spayed on crops during crop period. For the preparation of Jeevamruth we use cow urine and dung which contain numerous microorganisms. Further, we add forest soil or soil from anthill, as it contains diversified native soil microbes. These microbes multiply in the Jeevamruth due to fermentation process. So, by applying Jeevamruth we are inoculating microbes into the soil. Ghana Jeevamruth is the solid or powdered form of Jeevamruth. It is applied to soil before sowing.

24.3.3 Achandan

Achadan is mulching or covering of biomass on the soil surface. It should be done throughout the year. This supplies the biomass to the soil. Microbes utilize this biomass and supply nutrients to plant roots. Humus is generated by mulching process. Mulching is one of the management practices effectively used for maintaining soil moisture and for increasing soil organic matter. There are generally three types of mulching - soil, straw, and live mulching.

24.3.4 Waphasa

Waphasa is the soil aeration. It is that favourable microclimate in the soil created due to the mixture of air and water vapour.

24.4 IMPORTANCE OF COW BASED INPUTS

Though they have some quantities of nutrients, the application of cow based inputs is not meant for supply of nutrients but for the inoculation of microbial biota into the soil and further poly cropping system enhances the populations of microbes by synergistic actions with rhizosphere. It is proved that symbiotic microbes and a few species of free-living bacteria supply the nutrients to plants by consuming extrudates from plant root system. As long as nutrient uptake is facilitated, there will be no reduction in yields.

Cow based products and their importance in agriculture is well described by R.K. Pathak and R.A. Ram (2013). As per their review cow based products act as bioenhancers which stimulate the growth of microorganisms in the soil. Cow based products are rich sources of microbial consortia, macro-, micronutrients and plant growth promoting substances including immunity enhancers. They are also utilized to enhance decomposition of organic materials thereby enrich the soil and induce better plant vigour. The application of Ghanjeevamruth can stimulate and enhance the microbial population.

It is supported by the research work by Hoitink and Boehm (1999) who proved that microbial populations can be stimulated by the addition of organic amendments such as manure or compost.These preparations can be applied with irrigation water, drenched on organic mulches, diluted and sprayed as foliar application (Frank et al., 2005). These are known to supplement, major and minor nutrients act as growth stimulants and provide other beneficial substances to the plants (Sebastain and Christopher 2007).

24.5 365 DAYS GREEN COVER A NEW PARADIGM IN MAKING HEALTHY SOILS

In order to convert the soils into living soils or healthy soils within a short period, we need to maintain green cover throughout the year. Taking the advantages of Whapasa mechanism, established by natural farming practices, the plants absorb water even under drought conditions.

As per natural farming, there is no shortage of nutrients in the soil. But how these nutrients are made available to plant roots is important. This is done by the soil microorganism (bacteria, fungi, VAM) present on the roots and in the root zone. Plants utilize sunlight, water and CO_2 and convert into sugars (carbohydrates). 40% of sugars synthesized by plants are stored in the plant parts such as leaves and stem (aboveground parts). 30% of sugars are stored in the roots (below ground parts). And the remaining 30% plant sugars are extruded into the soil by roots. Microorganisms consume sugars, extruded by roots and in return they supply nutrients to roots in symbiotic manner. These microorganisms digest plant food material which are in the complex form and convert into mineral/elemental form and supply desired nutrients to plant roots. Unfortunately, this system is damaged by chemical farming. Soil microorganisms grow well under aerated moisture (Whapasa) conditions and utilize plant root exudates. This Whapasa is provided under natural farming conditions. The symbiotic and free-living bacteria and mycorrhizal hyphae supply nutrients to plants while enjoying Whapasa condition. Plants drawdown the atmospheric carbon into the soil very rapidly. So that the soil organic carbon becomes available within a short period. The mulching practice and continuous growth of plants enhances the soil organic matter. Under these favourable conditions, when Ghana and Drava Jeevamruth applied, soil microorganisms increase their diversity and number. On the whole the soil becomes healthy and ultimately turned as LIVING SOIL. The only thing is, we have to maintain roots to grow in the soil continuously. For this we should grow crops, cover crops, plants throughout the year, i.e., 365 days green cover on the land with diversity of crops.

24.6 HEALTHY LIVING SOIL PRODUCES HEALTHY CROPS

The soil resource is central to agriculture and therefore sustainable agriculture is inherently dependent on soil health. Many ecosystem processes have the soil as their regulatory centre and soil biota play a key role in a wide range of ecosystem services that underpin the sustainability of agro-ecosystems. The participatory development of soil health indicators and monitoring systems, integrating local and scientific knowledge, is proposed as a key component of a new approach, supporting farmers to adapt to agricultural intensification and attendant land-use and environmental change (Edmundo Barrios et al., 2014).

AP-ZBNF programme firmly believes that soil health is the key to agricultural productivity. In Andhra Pradesh, the community and agriculture department are together bringing the change in soil health by adopting ZBNF practices.

24.6.1 Humic Substances and Their Importance in Soil Fertility

Humic substances play a vital role in soil fertility and plant nutrition. Plants grown on soils which contain adequate humin, humic adds (HAs), and fulvic adds (FAs) are less subject to stress, are healthier, produce higher yields; and the nutritional quality of harvested foods and feeds are superior. The value of humic substances in soil fertility and plant nutrition relates to the many functions these complex organic compounds perform as a part of the life cycle on the earth. The life-death cycle involves a recycling of the carbon containing structural components of plants and animals through the soil and air and back into the living plant.

Humic substances provide energy and mineral requirements for soil microorganisms and soil dwelling animals. Beneficial soil organisms perform many beneficial functions which influence soil fertility and plant health. Beneficial bacteria release organic acids which are useful in solubilization of mineral elements of the soil. Bacteria also release complex polysaccharides that support in making of soil aggregates. Soil aggregates give soil to structure. Other beneficial soil microorganisms such as the Actinomyces release antibiotics into the soil. These antibiotics are taken up by the plant to protect it against pests. Antibiotics also function to create desirable ecological balance of soil organisms on the root surface (rhizoplane) and in soil near the roots (rhizosphere). Fungi perform beneficial functions in soils. Mycorrhizal fungi supply trace elements and water to plant roots. Other fungi decompose crop residues and vegetative organic matter and release nutrients. Many of the organic compounds released by fungi help in the formation of humus and soil aggregates. Beneficial soil dwelling animals create tunnel-like channels in the soil which are source for water percolation and air circulation. Soil dwelling animals also help in the formation of humus and balance the concentration of soil microorganisms within the food web. Thus, a living soil is a fertile and healthy soil.

The most important function of humic substances of soil is to hold water. Humic substances create a soil structure that facilitates water infiltration and water holding capacity for longer duration. Various carbon containing humic substances are important in the formation of soil aggregates. Complex carbohydrates synthesized by bacteria together with humic substances hold the clay and silt to form soil aggregates. These aggregates create a desirable soil structure in the top soil. Such soil has improved tilth and more porous openings. These pores facilitate gaseous interchange with the atmosphere and has greater moisture. This condition is called Waphasa in natural farming.

Because of air spaces and internal electrical charges, humic substances function as sponge. This sponge-like soil has the ability to hold six times more volume of water. Crops grown on such soil sponge can survive even in the drought periods. Water present in the soil sponge act as a carrier medium for nutrients required by soil organisms and plant roots. The healthy soil/soil sponge is created when 365 days green cover is maintained with application of mulch material and Ghana and Drava Jeevamruth.

Humic substances neutralize the pH of soil which enables absorption of more minerals and nutrients by plants. Humic substances promote healthy crop growth by enhancing uptake of major nutrients, viz., nitrogen, phosphorus, and potassium. Thus, when soil becomes fertile with the presence of adequate humic substances, requirement of N P K fertilizer is greatly reduced. That is why ZBNF farmers never apply NPK fertilizers for their crop growth, but maintain the healthy soil by incorporating humic substances. It is also proved that under the presence of humic substances, toxic metallic substances become unavailable to plants.The presence of humic substances in the soil influence the positive seeds germination and seedling development. They also act as root simulators. It is proved that seeds treated with Beejamruth and soils applied with Jeevamruth have resulted into more seed germination and root penetration in ZBNF fields.

24.6.2 Below Ground Biodiversity Induce the Immunity of Crops

AP-ZBNF programme is dedicated to conservation of biodiversity both below ground and above ground. Major focus is on below ground biodiversity by eliminating chemical inputs and enhancing organic matter and humic content. Below ground biodiversity is 10 times more than that of above ground biodiversity. Science has strong evidence that microbes such as Mycorrhizal fungi and plant growth promoting rhizobacteria (PGPR) help plants to deal with biotic and abiotic stresses through plant growth promotion and induced resistance. Below ground microbes interact with aboveground biodiversity such as plants (herbivore insects, their natural enemies and pollinators), in a bidirectional way.

The following research finding strongly support the ZBNF principles that healthy soils produce healthy plants, minimising the plant protection practices.

Schoonhoven L. M. et al. (2005) reported that plants have developed direct defence against insect herbivores and indirect defence to promote the effectiveness of natural enemies of insect herbivores. Soil-borne microbes (e.g., root endophytic fungi, mycorrhizal fungi, plant growth promoting fungi, rhizobacteria and rhizobia) exert positive effects on plant growth and survival through direct and plant mediated mechanisms. Plant defence mechanisms by plant growth promotion and induced systemic resistance (ISR) are described in detailed by Lugtenberg, B. and Kamilova, F. (2009), Van Loon, L.C. (2007), Schulz, B.J.E. (2006) and Trilllas, M.I. et al. (2009).

24.7 AGROECOSYSTEM THAT CONTROLS PESTS NATURALLY

AP-ZBNF cropping systems are based on agro-ecological principles. Regenerative and sustainable agriculture practices are designed on local and traditional knowledge. Various cropping systems, such as mixed cropping and intercropping with trap crops and boarder crops are commonly practised by farmers.

There is companion cropping system also where plant volatiles are used to manage key pests; both to repel pests and to attract beneficial organisms. This is called push-pull strategy. It is based on locally available plants, not expensive external inputs, and fits well with traditional mixed cropping systems in Africa (Khan et al., 2014).

Bund plantations (hedgerows) provide resources for pollinators by serving as host plants for insects and birds. Martins and Johnson, (2009) reported that farmers in the Kerio Valley plant and maintain highly diverse hedgerows that include both nectar forage plants and larval host plants for hawkmoths which are essential for the pollination of papaya.

Water conservation structures such as farm ponds and trenches in dryland areas with grass and small bushes serve to increase on-farm biodiversity and serve as nesting areas for bees. These areas also harbour spiders, dragonflies and praying mantises, and other natural enemies, all of which consume pest species (Martins, 2015).

Creating healthy ecosystems for the healthy crop growth and productivity is highly prioritized in ZBNF. All the practices that enhance crop growth through reducing weed competition and enhancing soil fertility for the growth of healthy crop which can tolerate weather externalities and pests. Few examples of such practices are: selection native varieties, application of Jeevamruth, minimum tillage, 365 days green cover, mulching with dead or live plant material, crop rotations, crop diversity, intercrops and mixed crops and relay cropping. These practices are supported by research findings of Derpsch and Friedrich, (2015). They have reported that farmers are practising conservation agriculture (similar to ZBNF) in East Africa for more than two decades and the number of farmers adopting this practice is increasing every year due to encouraging results.

Another agro-ecological use of ZBNF is augmentation of natural enemies in the local agro-ecosystems. Most of the natural enemies that have been lost during chemical farming period are to be bring back to manage the pest population naturally. Inter-cropping and boarder cropping, trap cropping and other crop or plant diversity maintenance is to provide the ecosystem for natural enemies. Below ground biodiversity, microorganisms have direct and indirect linkages with the defoliators and their natural enemies.

Crowder, D.W. et al. (2010) highlighted how communities of natural enemies affect pest populations.Understanding determinants for predator community composition and management of biological control services requires a landscape perspective. Both the pest and its enemies are often highly mobile and regulated at a much larger spatial scale than the level of the field, because they often require multiple resources, such as alternate food, hosts, and winter refuges to complete their life cycles (Tscharntke, T. et al., 2007 and Rusch, A. et al., 2010).

AP-ZBNF programme encourages farmers to use indigenous traditional knowledge (ITK) for pest and disease management, as prophylactic and control measures. Since thousands of years, use of botanicals, concoctions and cow based preparations are popular among farming community. List of various ITK products/concoctions are listed in the next chapter.

AP-ZBNF programme promotes crop diversification by introducing intercropping, boarder cropping, mixed cropping and relay cropping practices. AP-ZBNF main strategy is to promote intercropping so that farmers get cost of cultivation of main crop from the yields of intercrops. The recommended crops for intercrops are alternative crops of monocotyledons with dicotyledons crops. As per reports of NRC, (1993) and Tolera, (2003) this agro-ecological practice allows more efficient uses of on-farm resources. By growing more than one crop at a time in the same field, farmers maximize water use efficiency, maintain soil fertility, and minimize soil erosion, which are the serious drawbacks of mono-cropping (Hoshikawa, 1991).

Pest and disease incidence is greatly reduced in the polycropping models. No need to apply any external inputs. Under extreme cases, botanical extracts and other natural concoctions are advised. The following research supports our ongoing AP-ZBNF programme strategy to manage pest and diseases.

Mulumba et al. (2012), from Uganda reported that growing of different varieties of the same crop together consistently shows a decreased incidence of pest and diseases. Therefore, varietal diversity within a single crop system brings varying levels of resistance against the pests.

Intercropping of tomato with cowpea significantly reduced bacterial wilt when compared with tomato alone (Michal et al., 1997). Insect and diseases are less when tomato was intercropped with Maize (Pino et al., 1994). Gomez-Rodriguez et al. (2007) research report revealed that there was an effect of marigold intercropped with tomato crop. They found that intercropping with marigold induced a significant reduction in tomato early blight caused by *A. solani*, by means of three different mechanisms.

Another advantage of intercropping is the suppression of weed growth. Liebman and Dyck (1993) reported that weed population density and biomass production may be greatly reduced using intercropping (spatial diversification).

Intercrops may demonstrate weed control advantages over sole crops in two ways: (i) greater crop yield and less weed growth may be achieved if intercrops are more effective by suppressing weed growth through allelopathy. (ii) Intercrops may provide yield advantages without suppressing weed growth if intercrops use resources that are not exploitable by weeds or convert resources to harvestable material more efficiently than sole crops (Geno and Geno, 2001).

24.8 PLANT PROTECTION PRACTICES IN AP-ZBNF PROGRAMME

Even though healthy crops develop resistance against pest and diseases, under extreme conditions, some pests and diseases may pose threat to crops. To prevent any loss to the crop, the following non-negotiables are recommended for AP-ZBNF programme.

24.8.1 Seed and Seedling Treatment

Seed treatment with Beejamruth is non-negotiable in ZBNF. The following ingredients are used; water 20 lit, cowdung (desi cow) 5 kg, cow urine 5 lit, edible calcium 50 g and bund soil 200 g. This is practised to prevent seed-borne pests and diseases and also better germination.

24.8.2 Clipping of the Tips

Clipping of rice seedling tips at the time of transplantation to minimize carryover of rice Hispa, case worm and stem borer infestation from bed to the transplanted fields.

24.8.3 Alleys (For Paddy)

Leaving 20 cm feet path at every 2 metres interval in east-west direction in Kharif and north-south direction during Rabi season will avoid attack of hoppers and sheath blight.

24.8.4 White and Yellow Sticky Traps

Arrangement of 15-20 yellow and white sticky traps per acre is non-negotiable. Green leaf hoppers and thrips stick to these traps.

24.8.5 Bird Perches

Arrangement of 10-15 bird perches per acre is mandatory in paddy fields. This is to be done immediately after transplanting the paddy. Remove the perches at grain filling stage.

24.8.6 Pheromone Traps

Keeping 5-10 pheromone traps in zigzag way to mass trapping of adult males of different pests. Pheromone traps per acre above the plant height. Lure has to be changed once in a month or after the expiry date.

24.8.7 Azolla (For Paddy)

Application of Azolla at 15 days after transplantation, in Paddy fields, is non-negotiable. Apply 10 kg Azolla in to main field and within 15 days it will spread to the entire field.

24.8.8 Trap Crops

Growing of yellow flower marigold and castor in and around fields is non-negotiable. Need to ensure flowering of trap crops before main crop completes vegetative stage. This practice attracts pests to lay their eggs on the trap crops.

24.8.9 Boarder Crops

Growing of three rows of tall growing jowar or bajra or maize is non-negotiable. Boarder crops are to attract friendly insects and prevent pests and disease causing spores from neighbouring fields.

24.8.10 Light Traps

Light traps can be used to monitor and trap adult insects, thereby reducing their population. Some formal light traps that could be used are electric bulbs, hurricane lamps and bonfires. The adult moths, which get attracted by the bright light, fall into the water in the vessel and perish.

24.8.11 Application of Botanical Pesticides

If all the above-mentioned non-negotiables are followed religiously, there will be no need to apply botanical extracts. However, in case of severe infestation or to prevent the loss in advance, the use of botanical pesticides and natural concoctions is encouraged. The following botanicals and concoctions are recommended for plant protection. Farmers prepare at their own house or farm by collecting raw materials from cowshed and plant leaves from nearby sources like bund plantations, commons and forests.

1. Neemasthra
2. Agneyastra
3. Brahmasthra
4. Dashaparni concoction
5. Neem Paste
6. Vitex Concoction
7. Tobacco decoction.
8. Green chilli-garlic solution
9. Neem seed kernel extract (5%)
10. Cow dung- cow urine solution
11. Cow dung-cow urine-asafoetida solution.
12. Sour buttermilk solution
13. Dry ginger (shont)-Milk solution
14. Ipomoea solution
15. Sapthadhanyankura Kashay (Tonic)
16. Panchagavya (As tonic and plant protection)

REFERENCES

Bezemer, T.M. and van Dam, N.M. 2005. Linking aboveground and belowground interactions via induced plant defences. *Trends Ecol. Evol,* 20: 617–624.

Crowder, D.W. et al. 2010. Organic agriculture promotes evenness and natural pest control. *Nature* 466: 109–112.

Derpsch, R. & Friedrich, T. 2015. Global Overview of Conservation Agriculture Adoption. FAO www.fao.org/ag/ca/doc/derpsch-friedrich-global-overview-ca-adoption3.pdf

Edmundo Barrios, Keith Shepherd, Fergus Sinclair. Soil health and agricultural sustainability: The role of soil biota. In: Chapter 6 of Agroecology for Food Security and Nutrition Proceedings of the FAO International Symposium 18-19 September 2014, Rome, Italy.

Frank, E; Saro, G.R. and Radhakrishnan, M.T. 2005. Organic Cotton Crop Guide, a Manual for Practioners in the Tropics, FiBL, Switzerland, pp. 27-32.

Geno, L., and B. Geno, 2001. Polyculture Production: Principle, Benefits and Risk of Multiple Cropping. A report for the Rural Industry Research and Development Corporation (RIRDC), Publication, No. 01134.

Gomez-Rodriguez, O., E. Zavaleta-Mejia, A.V. Gonzalez-Hernandez, M. Livera-Munoz and E. Cardenas Soriano. 2007. Physiological and morphological adaptations in tomato intercropped with *Tageteserecta* and *Amaranthus hypochondriacus. Rev. Fitotec. Mex.,* 30(4): 421-428.

Hoitink HAJ, Boehm MJ. 1999. Biocontrol within the context of soil microbial communities: a substrate-dependent phenomenon. *Annu Rev Phytopathol,* 37: 427-446.

Hoshikawa, K. 1991. Significance of legumes crops in intercropping, the productivity and stability of cropping system. pp. 173-176. In: C. Johanson, K.K. Lee and K.L. Saharawat, (eds.). Phosphorus Nutrition of Grain Legume in the SemiArid Tropics, ICRISAT.

Khan, Z.R., Midega, C.A., Pittchar, J.O., Murage, A.W., Birkett, M.A., Bruce, T.J.A. & Pickett, J.A. 2014. Achieving food security for one million sub-Saharan African poor through push – pull innovation by 2020. Phil. *Trans. R. Soc. B,* 369: 20120284.

Liebman, M. and E. Dyck. 1993. Crop rotation and intercropping strategies for weed management. *Ecological Applications,* 3(1): 92-122.

Lugtenberg, B. and Kamilova, F. 2009. Plant growth-promoting rhizobacteria. *Annu. Rev. Microbiol,* 63: 541–556.

Martins, D.J. & Johnson, S.D. 2009. Distance and quality of natural habitat influence hawkmoth pollination of cultivated papaya. *International Journal of Tropical Insect Science*, 29(3): 114-123.

Martins, D.J. 2015. Passionfruit (Passiflora edulis) Farming in Kenya: Nathan Korir, a Farmer in the Kerio Valley in the North Rift. Kenya Pollination Project Case Study. FAO. In Press.

Michal, V.V., J.F. Wang, D.J. Midmore and G.L. 1997. Hartman. Effect of intercropping and soil amendment with urea and calcium oxide on the incidence of bacterial wilt of tomato and survival of soil borne *Pseudomonas solanacearum* in Taiwan. *Plant Pathology*, 46(4): 600-610.

Mulumba, J.W., Nankya, R., Adokorach, J., Kiwuka, C., Fadda, C., De Santis, P. & Jarvis, D.I. 2012. A risk-minimizing argument for traditional crop varietal diversity use to reduce pest and disease damage in agricultural ecosystems of Uganda. *Agriculture, Ecosystems and Environment*, 157: 70–86.

NRC (National Research Committee). Sustainable Agriculture and the Environment in the Humid Tropics. National Academy Press. Washington D.C., 1993, 702p.

Pathak R.K. and Ram R.A. 2013. Bio-enhancers: A potential tool to improve soil fertility, plant health in organic production of horticultural crops. *Progressive Horticulture*, 45(2): 237-254.

Pino, M., M.E. Domini, E. Terry, M. Bertoli and R. Espinos. 1994. Maize as a protective crop for tomato in conditions of environmental stress. *Cultivos Tropicales*, 15(2): 60-63.

Rusch, A. et al. 2010. Biological control of insect pests in agroecosystems: effects of crop management, farming systems, and seminatural habitats at the landscape scale: a review. *Adv. Agron.* 109: 219-260

Sebastain, P.S. and Christopher, L.A. 2007. Indigenous organic foliar sprays on crop yield. *Ecobiology*. 21: 201-207.

Schoonhoven, L.M. et al., eds. (2005). *Insect-Plant Biology*, Oxford University Press.

Schulz, B.J.E. (2006) Mutualistic interactions with fungal root endophytes. In: *Microbial Root Endophytes* (Schulz, B.J.E. et al., eds), 2005, pp. 261-279. Springer-Verlag, Berlin.

Tolera, A. 2003. Effects nitrogen, phosphorus farmyard manure and population of climbing bean on the performance of maize (*Zea mays* L.)/climbing bean (*Phaseolus vulgaris* L.) intercropping system in Alfisols of Bako. An MSc Thesis Presented to the School of Graduate Studies of Alemaya University. 1-75p.

Trillas, M.I. et al. 2009. Interactions between non-pathogenic fungi and plants. *Adv. Bot. Res.*, 51: 321-359

Tscharntke, T. et al. 2007. Conservation biological control and enemy diversity on a landscapescale. *Biol. Control,* 43: 294-309

Van Loon, L.C. 2007. Plant responses to plant growth-promoting rhizobacteria. *Eur. J. Plant Pathol,* 119: 243-254.

Van Wees, S.C.M. et al. 2008. Plant immune responses triggered by beneficial microbes. *Curr. Opin. Plant Biol,* 11: 443-448.

25 CHAPTER

Agro-Nanotechnology: An Eco-friendly Approach in the Management of Plant Disease and Mycotoxins

Halima Zugher Hussein

Department of Plant Protection, College of Agricultural Engineering Science, University of Baghdad, Iraq.

Corresponding Author: E-mail-drhalima@coagri.uobaghdad.eud.iq, halimaalbahadly@yahoo.com

ABSTRACT

Nanotechnology is a scientific field dealing with the understanding and study of material at a distance of about (1-100) nm as defined by the United States Environmental Protection Agency (USEPA), which includes building and restructuring physical, chemical, and biological properties of materials at the nanolevel, i.e., at the scale of atoms and molecules. In the twenty-first century, nanotechnology has emerged with a major impact on the global economy, industries, and public life. Historically, the agricultural nanotechnology researches were started since mid of twentieth century, but the real applications of this technology can be seen nowadays."Nano" is originally a Greek word means "a dwarf "and was used in mathematics to express the unit of length measurement (1 nanometre = 10^{-9} m, i.e., one part per billion of metre). It is used as a unit for measuring the length of small things that are only seen by an electron microscope, such as measuring the dimensions of atoms, molecules, and cells, as well as measuring the dimensions of microorganisms such as bacteria and viruses. Numerous studies have been conducted in recent years on the use of nanotechnology in agricultural production, which is called agro-nanotechnology. It is one of the most important tools in modern agriculture and includes employment of nanoparticles properties. Indeed, nano-molecules are absorbed efficiently by plants due to their small size that highly facilitates the penetration of the membrane of plant cells. Additionally, it enhances the crop productivity, protects the plant from insect pests and diseases, reduces agricultural crop losses and control of soil pollution due to heavy metals that impede plant nutrients and water absorption. Furthermore, it helps to purify water from organic and inorganic pollution. A great effort is being done to implement nanotechnology in food production, as it is believed that traditional agricultural systems will not have the capabilities to increase food crops. The rapid developments in nanotechnology have a major impact on agricultural practices, as some Nanomaterials have been implemented in plant diseases control, thus the mechanism for controlling plant diseases and reduction of mycotoxins with less side effects on crops or environment, which was the major concern of the researchers in the management of plant diseases. Therefore, it is necessary to search for alternative solutions and/or protocols to control plant diseases considering the environment safety. In recent years, the use of nanomaterials has expanded and become an alternative to control plant pathogens.

25.1 INTRODUCTION

Plant diseases are an essential aspect of human concerns, as their occurrence causes damage and loss to the plants that a humans depends on for their food, as the food shortage, may lead to starvation as it happening in some poor countries.The history has recorded some examples of several epidemic, as well as mass immigration, like the disaster of late blight of potatoes that was caused by *Phytophthora infestans,* led to the death of a quarter of a million Irish people as a result of the limited supply of potatoes and mass migration to America to escape starvation.Also, the blight disease on rice crops, which caused starvation in some Japanese islands–they were dependent on rice. Frost also caused the death of most of the citrus trees in Florida, the United States of America (USA), and the American Chestaal forests or oak were destroyed as a result of the blight caused by the *Cryphonctria parasitica.* Furthermore, the USA was also deprived of cultivating the finest European grapes *Vitis vinifera,* which is used in the production of the finest species Luxurious liquor due to the epidemiological infection of Pierces diseases caused by *Xylella fastidiasa,* besides other risks caused by plant diseases that influenced agriculture and industry. However, the diseases control the planting of some crops, due to the spread of some epidemic diseases, this leads farmers' reluctance from planting them, it also determines the type and the size of industry such as the production of pesticides or antibiotics to control plant diseases (Agrios, 2005). Fungal pathogens are produced what is known as the silent killer, which are mycotoxins the secondary metabolites produced by fungi under certain conditions. The most hazardous secondary metabolites that threaten human life are: T2-toxin, Zeralenone, Deoxynivanol, Fumonisine, Aflatoxin B1, Ochratoxin A. A lot of studies have proven that these materials can affecting vital systems by different ways, so they have got a great attention as natural contaminants (Hussein and Aizaldean. 2017). The mycotoxins constitute a large economic burden, as confirmed by many statistics available in a limited number of countries that are concerned with limiting the importance of mycotoxins and their economic impact, on the other hand there is absence of information about the pollution in most other countries around the world. The International Food and Agriculture Organization reports indicated that about 25% of world agricultural production is contaminated with mycotoxins. The infection with toxic fungi in agricultural production damages about a billion tonnes of agricultural production in the world annually. Moreover the effect of mycotoxins on human and animal health, the topic is difficult and very complicated due to the lack of accurate documented data and the lack of practical mechanisms for evaluation. Moreover, the effect of toxins on human and animal health is present and significant and may exceed the total other costs resulting from pollution with mycotoxins because all toxicological studies on various mycotoxins confirm the toxicity of these pollutants, in both acute and chronic toxicosis in addition to weaken the immune system (Hussein et al., 2011).For the above reasons, human thought to investigate the best way to control plant diseases and reduce them for safe life. Many approaches were examined to control mycotoxins including agricultural, physical, chemical, and biological methods, each has advantages and disadvantages. Recently, researchers have resorted to using environment-friendly materials that are effective in controlling plant pathogens, that involved the use of plant natural products. Nanotechnology has emerged as one of the most promising methods for controlling plant pathogens and mycotoxins contamination (Hussein, 2015). The application of nanotechnology in agriculture is relatively new compared to its applications in other fields. The US Department of Agriculture was the first to apply nanotechnology in agriculture and food industries in 2003, which led to the change of agricultural production systems. Nanotechnology applications improve production of food completely; therefore, the use of nanocapsules to cover the chemical materials, which are characterized by a high ability to solve many agricultural problems using nanotechnology in terms of controlling many agricultural pests.

25.2 THE MOST IMPORTANT SAFE METHODS FOR PRODUCING NANOMATERIALS

The production of the nanomaterials mostly accompanied with toxic side-products, the reason that forced scientists to resort to production of nanomaterials by other environment-friendly methods. Among these methods, the biosynthesis by fungi and bacteria and the green synthesis method by plants, as these biological systems work to produce mineral materials in nanoscale, gold and silver are examples of production nanomaterials by biological methods (Sadowski, 2010).

The most important advantages of biosynthetic nanomaterials are: (1) the low cost and high safety, (2) it produces nanoparticles with smaller dimensions than those produced by chemical, physical or mechanical methods, (3) the biosynthetic nanoparticles have catalytic efficacy and higher specific surface area, (4) they do not need stabilizers to prevent the aggregation and even if they aggregate, they'll not interfere with each other because of the presence of active cellular proteins such as cytochrome C, and the production is mostly higher with low consumed energy (Sheriff, 2010).

25.2.1 Biosynthesis of Nanomaterials by Bacteria

Bacteria are unicellular prokaryotic microorganisms with a simple structure and have no nucleus, among many genera of microorganisms, the bacteria took a greater interest in the field of biosynthesis of nanomaterials due to the ease of handling, and they can be modified genetically using genetic engineering techniques through expressing specific enzymes because the nucleus is lost. In recent times, interest in developing work to synthesis different nanomaterials using bacteria has increased (Mishra et al., 2015).Bacteriaare used in the production of nanomaterials due to their ability to reduction of mineral substances by special reduction enzymes such as reductases NADH dependent or nitro dependent reductase. Silver nanoparticles were produced by *Escherichia coli* (*E. coli*) which was spherical in shape with a size of 34.8 nm, and there is another type called magnetotatic bacteria such as *Aquaspirillum magnetotacticum*. Furthermore, *Magnetospirillum magneticum* have the ability to move along the magnetic field, which is a unique, and it forming nanoparticles inside cells with a size ranging from 1-120 nm (Sadowski, 2010).Whereas the bacterium *Pseudomonas stutzeri* AG 259 canforming silver nanoparticles around the plasm and the *Lactobacillus* can form silver and gold nanoparticles, as the gold nanoparticles were about 13 nm in size. *Enterobacter cloacae* can also synthesise silver nanoparticles, and the iron-reduction bacteria *Geobacter* can form iron nanoparticles, and recent research has indicated that *Acinetobacter baumannii* has the ability to synthesise silver nanoparticles with a size of 37 - 168 nm (Arshad, 2017, Shaker and Shabaan 2017).

25.2.2 Biosynthesis of Nanomaterials by Cyanobacterium

Cyanobacteric are unicellular to multicellular autotrophic bacteria, which can adapt to harsh environmental conditions and have the ability to produce nanoparticles as well as the ability to convert CO_2 to other forms of carbon after stimulation by sunlight. Silver and gold nanoparticles are the most nanoparticles that produced using *Cyanobacterium*, gold nanoparticles were synthesised inside cells using

Synechocystis sp. Pcc 6803, while *Plectonema boryanum* deposits gold nanoparticles on their cell walls when heated in a medium containing gold chloride (Mohanparia et al., 2008).

25.2.3 Biosynthesis of Nanomaterials by Plants

Green plants are living organisms belonging to the plant kingdom; they are multicellular and have the ability to reduct mineral ions faster than other biological systems. Generally, plants have been used to synthesise nanoparticles that can have advantages and are better than other biological systems such as the microscopic biological pathway, because they exclude preferably the processes of preserving the culture cell media. Additionally, they can be used in the mass production of nanoparticles because nanoparticles are produced by the biomass in plants then extracted from leaves, stems, flowers, or seeds. The mechanism of nanoparticles synthesis is represented by the presence natural reductants such as alkaloids, flavonoids, saponins, steroids, tannins and other nutrient compounds that act as reducing and stabilizing agents (Khan and Jamil, 2017). Silver nanoparticles were extracted from *Rosmarinus officinalis* leaf and *Eucalyptus chapmarriana* leaf, which proved to have a good antimicrobial activity (Sulaiman et al., 2015). Furthermore, silver nanoparticles from *Teucrium polium* extract have also demonstrated biological efficacy against bacteria, while the *Catharanthus roseus* produces silver nanoparticles in its leaves, which is used as antioxidants and wounds treatment. The *Heracleum persicum* plant proved effective in preparing 16 nm silver nanoparticles against *Acinetobacter baumannii* bacteria (Fajer, 2017)

25.2.4 Biosynthesis of Nanomaterials by Yeasts

Numerous studies have shown that yeasts can synthesise mineral nanoparticles, for example, it was indicated that the *Sacharomyces cervisiae* can absorb and reduce the Au^{+2} to the gold element in the peptidoglycan layer of their cell walls (within a living body) by the group of aldehydes present in the sugars. As well as it was pointed out that several different yeasts can form nano-gold in several shapes including spherical, triple and hexagonal within their cell body, especially in the cytoplasm.While the efficacy of three strains of *Rhodotorula glutinis, Saccharomyces cerevisiae,* and *Geotrichum candidum* were examined for biosynthesis of silver nanoparticles by direct exposure to the silver nitrate solution, and the size of the produced silver nanoparticles ranged between 2.5-20 nm. These are environment-friendly strains and it is well known as a high throughput nano silver producer and can be implemented in the traditional industry (Zahran et al., 2013).

25.2.5 Biosynthesis of Nanomaterials by Filamentous Fungi

The fungal synthesis of nanomaterials is a new intersection scientific field of mycology and nanotechnology that includes new sensational applications for interdisciplinary science with great potential due to its wide range and diversity of fungi.The use of fungi is potentially interesting because its high throughput enzymes production in addition to the simplicity of handling (for this reason *Streptomycetes* that have common features between fungi and bacteria, have high potential for the production of nanoparticles).The fungal systems or myco-nano factories have actually exploited to synthesise nanostructures in silver, gold, zirconium, silica, titanium, iron, etc. Fungi attracted the attention of many researchers concerning the production of bio-metallic nanomaterials due to their ability to tolerate and bioaccumulation of minerals, the extracellular enzyme secretion, which facilitates massive production of enzymes (Longoria et al., 2012). The economic ability and ease of development of the biomass of fungi is another advantage of using fungi in the safe production of metal nanoparticles, as several species can grow rapidly so the culture and preservation of fungi in the laboratory is easy. Furthermore, the fungi can produce metal nanomaterials through reduction enzymes outside or inside the cell. The synthesis of silver particles by fungi is a method of bottom-up methods, as the primary reaction is either a reduction or oxidation of the base material due to the increase in colloidal materials. This proves that fungal enzymes are responsible for reducing mineral compounds to nanoparticles; other advantage of bottom-up method is the obtaining of nanoparticles with chemical homogeneity and low imperfections (Zielonka and Klimek, 2017). Silver nanoparticles were produced from a wide range of fungal species, such as *Penicillium expansum*, which produced spherical silver nanoparticles with a size of 20 nm (Dubey et al., 2012). *P. expansum* HA2N produced nanoparticles with a size ranging from 14-25 nm that showed efficacy as an antimicrobial agent, also oval silver nanoparticles were produced by *Rhizoctonia* sp. with a size ranging from 25 to 50 nm. The *A spergilli* were examined for their ability to produce nanoparticles of and the results revealed that *A. niger*, had the ability to produce nanoparticles ranged between 20-55 nm in size (Ninganagouda et al., 2013), while *A. terrus* HAIN, produced particles were ranged between 10-18 nm, which proved effective in reduction the Ochratox in (Ammar and El-Desouky, 2016). Furthermore, *P. chrysogenum* and other *Penicilli* produced nanoparticles were used as excellent antimicrobials, also *Fusarium oxysporum* produced silver nanoparticles of 5-13 nm in size with spherical shape (Husseiny et al., 2015). Likewise, *Arthroderma fulrum* produces silver nanoparticles sized 13.5-17.5 nm with a spherical shape. Additionally, *Raphanns satirum* showed possible ability to produce silver nanoparticles ranging between 4-30 nm with a spherical shape (Xue et al., 2016).

25.3 NANOTECHNOLOGY: THE SPECIAL CONCEPTS AND THEIR ROLE IN THE CONTROL OF PLANT DISEASE AND MYCOTOXINS

25.3.1 The Special Concept of Nanotechnology

The European Union defines the geometric nanomaterials as intentionally processed materials that have one or more dimensions equals or less than 100 nm, or aggregated to compose granules provided that they retain their nanoscale properties. The properties of the nanomaterials generally include the large surface area with several physical and chemical properties that differ from the mother material (Peters et al., 2014). Nanomaterials have become the cornerstone of nanoscience and nanotechnology, as the technology of nano-construction is a broad and interdisciplinary field of research that developed significantly all over the world in the past few years. Moreover, it has the potential to revolutionize how materials and products are created, the scope and nature of the jobs that can be accessed; it does have a huge commercial impact, which will increase in the future (Alagarasi, 2013). Nanomaterials are classified into inorganic minerals, inorganic semiconductors, and inorganic insulators as well as inorganic nanomaterials and polymers. There are other classifications based on chemical, physical, biological origin, or based on their gaseous, liquid, solid, synthesis, or based on energy source, laser, microwave, acoustic chemistry and other optical, mechanical, and thermal properties (Bensebaa, 2013).

25.3.2 The Application in Controling Plant Diseases and Toxins

Ying Hua et al. (2005) stated that nanolayer was developed based on montmoronite to absorb mycotoxins from the animal's digestive system to protect them from poisoning. Shi et al. (2006) used modified montmoronite to reduce the toxicity of aflatoxin in chickens.Nano chitosanis considered a promising agent for protecting foods due to its efficiency in eliminating microorganisms, its high ability to absorb toxins, high susceptibility to biodegradation, non-toxic, and its advanced chemical and physical properties (Dutta et al., 2009). Recent trends in the elimination of mycotoxins from food and feed have led to the use of clay particles such as bentonite and zeolite because of their opposite polarity as absorbents for mycotoxins (Dal Pozzo et al., 2016). Brown et al. (2014) and Bhatti et al. (2018) stated that these methods have disadvantages due to their ability to bind minerals and vitamins of the feed, therefore, when apply these substances, the addition of minerals and vitamins should be increased by 20%. Carbon nanomaterials have been used as absorbents to reduce mycotoxins and they showed a high adsorption capacity such as grapheme oxide GO. These substances also play an important role in interactions with organic compounds, in addition to their ability to absorb aromatic rings from these compounds, where it was used in reduction of the toxins from herbal medicines (Wang et al., 2014; Zhang et al., 2018). Hussein et al. (2017) confirmed that the MgONPs and the ZnNPs inhibit the growth of *Pythium aphanidermatum* on cucurbits by 98% and 100% of the MgONPs and 97, 99.80% of the ZnNPs. Moreover, Yousif et al. (2019) reported that the use of ZnONPs had decreased the concentration of Aflatoxin B1 (AFB1) toxicity by 93.8%, while honey achieved a 79.64% reduction ratio. As for using honey supported with ZnONPs, it achieved 82.3% reduction in toxin when these materials were used with stored maize. Al Rawi and Hussein, (2017) confirmed that the silver nanoparticles (AgNPs) showed high efficiency in inhibiting the growth of *A. flavus*, as the fungi were 100% inhibited when using the 175 ppm and 200 ppm concentration, while the 150 ppm concentration caused 95% inhibition. Abdul-Hasan and Hussein (2016) evaluated the effect of silver oxide nanoparticles, AgNPs, and nano magnesium oxide MgNPs and they confirmed that the growth of the *Fusarium Solani* f. sp *cucurbitae* on musk melon, was inhibited 100%, the treatments did not affect dry and fresh weight and productivity in greenhouses. Al-Qaisi et al. (2015) achieved a 100% inhibition of *A. flavus* using a mixture of both MgONPs and SiO_2NPs with a concentration of 3% that causeda reduction in the AFB1 toxicity by 100%. As well as, the same mixture achieved 34.71% of geowth inhibition and 13.71% of AFB1 reduction ratio for when examined on maize. It has been confirmed that silver nanoparticles are one of the effective nanomaterials in eliminating mycotoxigenic fungi growth and toxin production. The big challenge of most physical and chemical production methods is the expensive coast and employment of hazardous materials that may constitute a real danger, therefore there is a need to synthesize them biologically. Kalishwaralal et al. (2008) and Al Rawi (2017) found that addition of silver nanoparticles to the stored maize at a concentration of 175 ppm prevented the *Aspergillus flavus* from producing AFB1 and the same concentration decreased the toxin concentration to 0.1 μg/g maize grain compared to 0.6 μg/g maize in control treatment. Gardea et al. (2003) synthesized gold and silver nanoparticles using alfalfa plant for the first time, given that *Cannabis* contains many chemicals including antibiotics that have similar properties to penicillin. Bala et al. (2013) achieved biosynthesis of silver nanoparticles using the water extract of *A. fumigatus* isolated from the *Cannabis sativa* leaves, and these particles have anti-bacterial activity. Al-Juburi (2016) used ZnOPs to stimulate peanut resistance against *Aspergillus flavus*. While Al-Joudary, (2017) used both iron and zinc nanoparticles to improve the properties of *Calotropis procera* plant. Furthermore, Hussain (2017) used MgONPs as inhibitor to *Fusarium solani* on musk melon, Al Khlefawy (2017) used nanochelated iron to improve the efficiency of Moringa plant, while Zaidi, (2018) used zinc nanoparticles (ZnONPs) to improve cauliflower's efficacy against botulinum disease. Marzouq (2018) supported honey with nano-magnesium oxide to control *Aspergillus flavus*, which infects maize seeds under storage conditions, also Fawzy (2018) used MgONPs to inhibit the same fungus. Al-Dulaimi (2019) used MgONPs to induce systemic resistance of tomato plants against *F. oxysporumf* sp. *lycopersici*. In addition, Taha (2019) worked on the synthesis of silver nanoparticles using *Penicillum italicum* and evaluated its efficiency, whereas Al-Tamimi (2019) used ZnOPs to inhibit the *Rhizocto niasolani* on tomato plants. Dakel et al. (2020) demonstrated the efficacy of nanoalgae (*Ascophylim*sp.) in protecting wheat from root rot disease, which induced the resistance in plants (unpublished research). Hussein et al. (2020) discovered the activity of MgONPs as inhibitor to the *Didymell abroni in vitro* as well as it reduced the severity of infection *in vivo* (unpublished research). Ibrahium et al. (2020) demonstrated the efficacy of nanochitosan in induction of systemic resistance in tomato plant against *Alternaria solani* (unpublished research). Additionally, Tahr et al. (2020) determined the efficiency of both *Citrus aurantium* L. seeds and normal and sea salt nanoparticles in inhibiting *Aspergillus flavus* and reduction AFB1 (unpublished research). Hekal et al. (2020) confirmed the efficacy of nanochitosan, urea and fylaxin inhibition the *Aspergillus flavus* growth and reduction AFB1 (unpublished research). Salam, (2020) showed the efficacy of the *Gunderma* sp. in inhibition of *Aspergillus flavus* growth and reduction of AFB1 (unpublished research). Hamoudi et al. (2020) demonstrated the efficiency of the nanoalgae (*Ascophylum* sp.) extract in inhibition of *Seclerotenia* growth at a concentration of 2% in vivo. It confirmed that it has a positive effect in controlling white mould disease in pot experiments, and the induction of systemic resistance in the eggplant by increasing the activity of the peroxidase enzyme and some nutrients in the plant (unpublished research). Despite the prevalence of the metal nanoparticles synthesis by bacteria, fungi have been used in the biosynthesis of nanomaterials, because they are more beneficial since mycelium provides a large area for interaction. As well as due to the fungi secretion of relatively larger amounts of proteins, which made the mineral salts to convert to metal nanoparticles faster (Pantidos and Horsfall, 2015; Siddiqi and Husen, 2016).Kovač et al. (2018) used three concentrations of Fullerene C60 (Fullerene C60, which is soluble in water and transforms into fullerenol 24C60 (OH), has great potential in modifying the synthesis of secondary metabolism of fungi, to demonstrate this potential, the product was used to inhibit the biosynthesis of the aflatoxin producing precursors produced by *A. flavus*) 10, 100 and 1000 ng/mL in yeast extract medium against *A. flavus* during 7 days at 29 °C, the mycelium growth was affected at 1000 ng/mL, while 10 ng/mL^{-1} influenced the formation of conidia, and all concentrations affected the production of aflatoxin by regulation of signal transmission and gene expression in the early stages of the synthesis, as well as impeding the activity of one of vital enzymes to produce aflatoxin. The nanoparticles that involve functional groups can interact with biological molecules such as peptides, nucleic acids and proteins have been developed and used in biosensors to detect and amplify different signals (Bönnemann and Richards, 2001). Magnetic nanoparticles were used to stabilize the movement of graphene oxide and were used as biological receptors to adsorb the aflatoxin M and B molecules (Rai et al., 2012; Gan et al., 2013). Bonel et al. (2010) mentioned that the biosensors showed great progress in analyzing mycotoxins and were able to measure the very low concentrations at the work site by using small and fast tools and when adding improvements to the sensor device with gold nanoparticles, they became more sensitive in detection of up to 1.5 ng/mL. Furthermore, Moon and Lee (2013) developed sensors in the form of slides that are very sensitive, less expensive and smaller, and can be used to analyze mycotoxins in different stages of food from the field to storage to stores, therefore they can be considered fast detectors to perform qualitative and quantitative examination of mycotoxins. Campagnoli et al. (2011) used the electronic nose as a sensor to detect mycotoxins in naturally contaminated wheat, as well as used as an absorbent material to distinguish contaminated samples from healthy wheat grains. Other nanomaterials were also used in sensors, such as silver nanoparticles, to detect the Ochratoxinin beer as they detected concentration of 0.02 μg/ L (Evtugyn et al., 2013). Quantum dots were also used with the aptamers as sensors to detect the Ochratoxinin wine, where within 10 minutes they detected 1.9 μg / L (Wang et al., 2011). Wang et al. (2015) used SiO_2 NPs to design a sensor can detect Ochratoxinin peanuts, which was able to detect 0.0005 and 0.012 μg/L. This strategy may provide fast and highly sensitive devices to examine mycotoxins contamination. Gold nanoparticles sensors were used as a chromatic probe to detect AFB1 and AFB2, the readings were seen by the eye, the first tested sensor was able to detect 0.025 μg/L of both AFB1 and AFB2 (Luan et al., 2015). A sensor was developed by SERS Company to double detect the toxicity of AFB1 and OTA in corn at concentrations of 0.006 and 0.03 μg/L (Zhao et al., 2015).

REFERENCES

Agrios, G.N. 2005. "Plant Pathology." 5th Edition: Elsevier Academic Press. Burlington, Ma. USA. 79-103.

Abdul-Hasan, F. and Hussein, H.Z. 2016. "Evaluation of the antifungal effect of nanoparticles (Magnesium oxide, silver) and chemical (Phylax) on *Fusarium solani* f. sp. *Cucurbitae,* pathogenic agent of melon". *International Journal of Agriculture Innovations and Research.* 5(2): 276-282.

Al-Qaisi, I.A.A. 2015. "Efficiency test of some MgO and SiO_2 nanomaterials in reducing Aflatoxin B1 in stored yellow maize." PhD thesis. Faculty of Agriculture – University of Baghdad.

Al-Rawi, M.A. 2017. "Diagnosis of the gene responsible for the production of Aflatoxin B1 in *Aspergillus flavus* and to reduce it by silver nanoparticles". Master Thesis. *College of Agriculture, Univ., of Baghdad.*

Al-Rawi, M.A. and Hussein, H.Z. 2017. "Antifungal effect of silver nanoparticles (AgNPs) Against *Aspergillus flavus*". *EC Microbiology* 6(3): 63-66.

Ammar, H.A.M., and T.A. 2016. El-Desouky. "Green synthesis of nanosilver particles by Aspergillus terreus HA 1N and *Penicilliumexpansum* HA 2N and its antifungal activity against mycotoxigenic fungi." *Journal of Applied Microbiology,* 121(1): 89-100.

Alagarasi, A. 2013. "Introduction to nanomaterials." *Indian Institute of Technology Madras,* 1-24.

Arshad, Ayesha. 2017. "Bacterial synthesis and applications of nanoparticles." *Nano Sci. Nano Technol.* 11(2): 119.

Bala, Manju, and Vedpriya Arya. 2013. "Biological synthesis of silver nanoparticles from aqueous extract of endophytic fungus *Aspergillus fumigatus* and its antibacterial action." *International Journal of Nanomaterials and Biostructures* 3(2): 37-41.

Basaran, Pervin, Nese Basaran-Akgul, and LutfiOksuz. 2008. "Elimination of *Aspergillus parasiticus* from nut surface with low pressure cold plasma (LPCP) treatment." *Food Microbiology,* 25(4): 626-632.

Bhatti, Sheraz Ahmed, Muhammad Zargham Khan, ZahoorUl Hassan, Muhammad Kashif Saleemi, Muhammad Saqib, Aisha Khatoon, and Masood Akhter. 2018. "Comparative efficacy of Bentonite clay, activated charcoal and *Trichosporon mycotoxinivorans* in regulating the feed-to-tissue transfer of mycotoxins." *Journal of the Science of Food and Agriculture,* 98(3): 884-890.

Bonel, L., Juan C. Vidal, P. Duato, and Juan R. Castillo. 2010. "Ochratoxin A nanostructured electrochemical immunosensors based on polyclonal antibodies and gold nanoparticles coupled to the antigen." *Analytical Methods,* 2(4): 335-341.

Brown, K.A., T. Mays, A. Romoser, A. Marroquin-Cardona, N.J. Mitchell, S.E. Elmore, and T.D. Phillips. 2014. "Modified hydra bioassay to evaluate the toxicity of multiple mycotoxins and predict the detoxification efficacy of a clay-based sorbent." *Journal of Applied Toxicology,* 34(1): 40-48.

Campagnoli, Anna, Federica Cheli, Carlo Polidori, Mauro Zaninelli, OresteZecca, Giovanni Savoini, Luciano Pinotti, and Vittorio Dell'Orto. 2011. "Use of the electronic nose as a screening tool for the recognition of durum wheat naturally contaminated by deoxynivalenol: a preliminary approach." *Sensors,* 11(5): 4899-4916.

Dal Pozzo, Marcelo, Julio Viegas, Gilberto VilmarKozloski, Cristiano Miguel Stefanello. 2016. Alisson Minozzo da Silveira, Cimélio Bayer, and JanioMoraisSanturio, "The effect of mycotoxins adsorbents beta glycan or montmorillonite on bovine ruminal fermentation in vitro." *Acta Scientiae Veterinariae,* 44(1): 6.

Dubey, M., S. Sharma, S. Bhadauria, R.K. Gautam, and V. M. Katoch. 2012. "Fungal Biosynthesis of Antimicrobial Nanosilver Solution: A Green Approach." In: *Chemistry of Phytopotentials: Health, Energy and Environmental Perspectives*, pp. 53-57. Springer, Berlin, Heidelberg.

Evtugyn, Gennady, Anna Porfireva, Rusal Sitdikov, Vladimir Evtugyn, Ivan Stoikov, Igor Antipin, and Tibor Hianik. 2013. "Electrochemical aptasensor for the determination of ochratoxin A at the Au electrode modified with Ag nanoparticles decorated with macrocyclic ligand." *Electroanalysis,* 25(8): 1847-1854.

Gand F.D.L. 2003. "Study the synergistic effect of silver nanoparticles with antibody against isolates of *Acinetobacter baumannii* isolated from patients with skin lesions". Master thesis. *College of Basic Education, Department of Science, Al-Mustansiriya University.*

Gan, Ning, Jing Zhou, Ping Xiong, Futao Hu, Yuting Cao, Tianhua Li, and Qianli Jiang. 2013. "An ultrasensitive electrochemiluminescent immunoassay for Aflatoxin M1 in milk, based on extraction by magnetic graphene and detection by antibody-labelled CdTequantumn dots-carbon nanotubes nanocomposite." *Toxins,* 5(5): 865-883.

Gardea-Torresdey, Jorge L., Eduardo Gomez, Jose R. Peralta-Videa, Jason G. Parsons, HoracioTroiani, and Miguel Jose-Yacaman.

2003. "Alfalfa sprouts: a natural source for the synthesis of silver nanoparticles." *Langmuir*, 19(4): 1357-1361.

Hassan, F.F. and Hussein, H.Z. 2017. "Detection of aflatoxin M1 in pasteurized canned milk and using of UV radiation for detoxification". *Int'l Journal of Advances in Chemical Eng., and Biological Sciences*, 4(1):130-133.

Hussein, H.Z., D.L. Aiz-aldean and O.N. Manty. 2011. "The first national symposium on mycotoxins and their environmental hazards". Univ. of Baghdad, 111 pp.

Hussein, Halima Z. 2015. "Activity of pomegranate peels and clove powders in detoxification of aflatoxin B1 and ochratoxin A from contaminated poultry diet." *Journal of Plant Pathology and Microbiology*, 6: 1.

Hussein, H.Z., and D.L. Aiz-aldean. 2017. "Managing plant diseases and mycotoxins between a glorious past, a bright present and a promising future". Univ. of Baghdad, 176 pp.

Hussein, Halima Z. 2017. "Detection of trichothecene of *Fusarium solani* isolates by using HPLC in melon plants." *Pak. J. Biotechnol*, 14: 211-213.

Hussein, Halima Z., Eman K. Abdul-Karim, and Shamael S. Mutar. 2017. "Possibility of using nanoparticles (ZnNPs, MgONPs) in keeping cucurbit fruit from infection by *Pythium aphanidermatum*." *Int J Sci Res*, 6(5): 837-839.

Hussein, Halima Z., Luma K. Bandr, and F.M. Alkhilani. 2017. "Testing the efficiency of probiotic in reducing the growth of *Aspergillus flavas* in corn stored and its effect on performance of broiler." *Research Journal of Pharmaceutical Biological and Chemical Sciences*, 8(3): 1791-1801.

Hussein, Halima Z., Riyam H. Tuama, and Amna M. Ali. 2015. "Study the effect of ozone gas and ultraviolet radiation and microwave on the degradation of aflatoxin B1 produce by *Aspergillus flavus* on stored Maize grains." *IOSR J. Agri. Vet. Scie*, 8(5): 5-12.

Keiper, Adam. 2003. "The nanotechnology revolution." *The New Atlantis*, 2: 17-34.

Kovač, Tihomir, Ivana Borišev, BiljanaCrevar, FraneČačićKenjerić, MarijaKovač, Ivica Strelec, Chibundu N. Ezekiel, Michael Sulyok, Rudolf Krska, and BojanŠarkanj. 2018. "Fullerol C 60 (OH) 24 nanoparticles modulate aflatoxin B 1 biosynthesis in *Aspergillus flavus*." *Scientific Reports*, 8(1): 1-8.

Luan, Yunxia, Jiayi Chen, Gang Xie, Cheng Li, Hua Ping, Zhihong Ma, and Anxiang Lu. 2015. "Visual and microplate detection of aflatoxin

B2 based on NaCl-induced aggregation of aptamer-modified gold nanoparticles." *Microchimica Acta,* 18(5-6): 995-1001.

Luan, Yunxia, Zhengbo Chen, Gang Xie, Jiayi Chen, Anxiang Lu, Cheng Li, Hailong Fu, Zhihong Ma, and Jihua Wang. 2015. "Rapid visual detection of aflatoxin B1 by label-free aptasensor using unmodified gold nanoparticles." *Journal of Nanoscience and Nanotechnology,* 15(2): 1357-1361.

Mishra, Amrita, Suraj Kumar Tripathy, Rizwan Wahab, Song-HoonJeong, Inho Hwang, You-Bing Yang, Young-Soon Kim, Hyung-Shik Shin, and Soon-Il Yun. 2011. "Microbial synthesis of gold nanoparticles using the fungus *Penicillium brevicompactum* and their cytotoxic effects against mouse mayo blast cancer C 2 C 12 cells." *Applied Microbiology and Biotechnology,* 92(3): 617-630.

Mohanpuria, Prashant, Nisha K. Rana, and Sudesh Kumar Yadav. 2008. "Biosynthesis of nanoparticles: technological concepts and future applications." *Journal of Nanoparticle Research,* 10(3): 507-517.

Moon, Jihea, Giyoung Kim, and Sangdae Lee. 2013. "Development of nanogold-based lateral flow immunoassay for the detection of ochratoxin A in buffer systems." *Journal of Nanoscience and Nanotechnology,* 13(11): 7245-7249.

Pantidos, Nikolaos, and Louise E. Horsfall. 2014. "Biological synthesis of metallic nanoparticles by bacteria, fungi and plants." *Journal of Nanomedicine and Nanotechnology,* 5(5): 1.

Rai, Mahendra K., Shital R. Bonde, Avinash P. Ingle, and Aniket K. Gade. 2012. "Mycotoxin: rapid detection, differentiation and safety." *Journal of Pharmaceutical Education and Research,* 3(1): 22-34.

Peters, Ruud, Puck Brandhoff, Stefan Weigel, Hans Marvin, Hans Bouwmeester, Karin Aschberger, Hubert Rauscher et al. 2014. "Inventory of Nanotechnology applications in the agricultural, feed and food sector." *EFSA Supporting Publications,* 11(7): 621E.

Rhouati, Amina, Gonca Bulbul, Usman Latif, Akhtar Hayat, Zhan-Hong Li, and Jean Louis Marty. 2017. "Nano-aptasensing in mycotoxin analysis: Recent updates and progress." *Toxins,* 9(11): 349.

Sadowski, Zygmunt. 2010. "Biosynthesis and application of silver and gold nanoparticles." *Silver Nanoparticles.* 13: 257-276.

Shaker, Mohamed A., and Mona I. Shaaban. 2017. "Synthesis of silver nanoparticles with antimicrobial and anti-adherence activities against multidrug-resistant isolates from *Acinetobacter baumannii.*" *Journal of Taibah University Medical Sciences,* 12(4): 291-297.

Shi, Y.H., Z.R. Xu, J.L. Feng, and C.Z. Wang. 2006. "Efficacy of modified montmorillonite nanocomposite to reduce the toxicity of aflatoxin in broiler chicks." *Animal Feed Science and Technology,* 129(1-2): 138-148.

Siddiqi, KhwajaSalahuddin, and AzamalHusen. 2016. "Fabrication of metal nanoparticles from fungi and metal salts: scope and application." *Nanoscale Research Letters,* 11(1): 98.

Sulaiman, Ghassan M., H.I.B.A.T Hussien, and Maysoon MNM Saleem. 2015. "Biosynthesis of silver nanoparticles synthesized by *Aspergillus flavus* and their antioxidant, antimicrobial and cytotoxicity properties." *Bulletin of Materials Science,* 38(3): 639-644.

Taha, Z.K. 2019. "Biosynthesis of silver nanoparticles by fungus *Penicillum italicum* and evaluation, antioxidant, antimicrobial, and anticancer activity". A thesis, College of Education for Pure Science, Univ. of Baghdad. 82 pp.

Wang, Chengquan, Jing Qian, Kun Wang, Mengjuan Hua, Qian Liu, Nan Hao, Tianyan You, and Xingyi Huang. 2015. "Nitrogen-doped graphene quantum dots@ SiO_2 nanoparticles as electrochemiluminescence and fluorescence signal indicators for magnetically controlled aptasensor with dual detection channels." *ACS Applied Materials and Interfaces,* 7(48): 26865-26873.

Wang, Libing, Wei Chen, Wenwei Ma, Liqinag Liu, Wei Ma, Yuan Zhao, Yingyue Zhu, Liguang Xu, Hua Kuang, and Chuanlai Xu. 2011. "Fluorescent strip sensor for rapid determination of toxins." *Chemical Communications,* 47(5): 1574-1576.

Xue, Baiji, Dan He, Song Gao, Dongyang Wang, Koji Yokoyama, and Li Wang. 2016. "Biosynthesis of silver nanoparticles by the fungus *Arthroderma fulvum* and its antifungal activity against genera of Candida, Aspergillus and Fusarium." *International Journal of Nanomedicine,* 11: 1899.

Yousif, D.A., Hussein, H.Z. and Ali, A.M. (2019).Testing the efficiency of zinc oxide nanoparticles, honey, honey supported by zinc oxide nanoparticles and phylex in degradation of aflatoxin B1 in stored sea maize seeds. *Journal of Global Pharma Technology.* 11(2): 225-33.

Yinghua, Shi, Xu Zirong, and Feng Jianlei. 2005. "In vitro adsorption of aflatoxin adsorbing nano-additive for aflatoxin B1, B2, G1, G2." *Scientia Agricultura Sinica.* 38(5): 1069-1072.

Zahran, M.K., A.A. Mohamed, F.M. Mohamed, and M.H. El-Rafie. 2013. "Optimization of biological synthesis of silver nanoparticles by some yeast fungi." *Egypt J Chem,* 56(1): 91-110.

Zhang, Lei, Xiao-Wen Dou, Cheng Zhang, Antonio F. Logrieco, and Mei-Hua Yang. 2018. "A review of current methods for analysis of mycotoxins in herbal medicines." *Toxins,* 10(2): 65.

Zhao, Yuan, Yaxin Yang, Yaodong Luo, Xuan Yang, Manli Li, and Qijun Song. 2015. "Double detection of mycotoxins based on SERS labels embedded Ag@ Au core–shell nanoparticles." *ACS Applied Materials and Interfaces,* 7(39): 21780-21786.

Zielonka, Aleksandra, and Magdalena Klimek-Ochab. 2017. "Fungal synthesis of size-defined nanoparticles." *Advances in Natural Sciences: Nanoscience and Nanotechnology,* 8(4): 04301.

26 CHAPTER Management of Fungal Diseases of Crops Through Plant-Based Natural Products

Enespa[1] and Prem Chandra[2]
1. Department of Plant Pathology, School for Agriculture, Sri Mahesh Prasad Degree College, University of Lucknow, Lucknow- 226007
2. Department of Environmental Microbiology, School for Environmental Sciences, BBAU, Lucknow, India- 226025

Corresponding author: E-mail: enespasingh@gmail.com

ABSTRACT

Most of the phytopathogenic diseases caused by the fungi that they have gigantic reproductive capability and hold the metabolic mechanisms which are able to cultivate resistance to marketable fungicides. The mycotoxins on crops produce phytopathogenic fungi *Aspergillus, Alternaria, Botrytis, Colletotrichum, Fusarium, Phytophthora, Penicillium* and *Pythium,* among others. Several plants belonging to the families of Anacardiaceae, Asteraceae, Caprifoliaceae, Lamiaceae, Liliaceae, Papilionacea, Poaceae, Verbenaceae are gorgeous sources of bioactive secondary metabolites of wide spread diversity such as tannins, terpenoids, saponins, alkaloids, flavonoids, essential oils and other compounds, reported to have in vitro antifungal properties. So, the secondary metabolites with antifungal activity signify an alternative for succeeding a sustainable control of phytopathogenic fungi and to reduce the heavy reliance of synthetic pesticides used to control them.

26.1 INTRODUCTION

The substantially subsidized to overall loss in revenue among crop plants by phytopathogenic fungi which are one of the major biotic stresses (Vurro et al., 2010). Phytopathogenic fungi ruine crops and staple foods approximately 10-20%. For this fungus the control policies have been established on the use of artificial fungicides (Engelmeier & Hadacek, 2006). However, due to the attendance of highly impervious isolates this method has become less effective and more constrained and to the adverse environmental effects that originate from the unselective use of these products (Chithra et al., 2015). For the promotion of sustainable and vigorous yields of foods the naturally safe agrochemicals produced by the unconventional agriculture systems which motivates the improvement of research concerned to provide alternative mechanism policies, being the use of plants as a foundation of secondary metabolites with antifungal possessions (García-Mier et al., 2013). A wide range of biological activities possessed by the secondary metabolites of plants lead to interactions of diversity with other organisms (Lattanzio et al., 2008). The natural products of plants offer a pool of structurally varied antifungal compounds in this reference, to control pathogenic fungi of plants providing an alternative to artificial fungicides (Ribera & Zuñiga, 2012). These compounds, active against a restricted number of particular target species, are suitable potentially for their use as agrochemicals and biodegradable in pest management programmes jointly (Khater, 2012). To obtain higher yields in agriculture from competing pathogens the crop protection management is necessary. To use eco-friendly approaches for controlling agriculture production is currently trending. To search the novel, affordable, and less toxic methods for pest control led to the growing request of pesticide-free agricultural products (Saini et al., 2020). The chemical compounds formed from natural products are active and that could interrupt or inhibit pathogen growth (Dayan & Duke, 2014). Throughout their development to prevent contaminations due to pathogens; a large number of secondary metabolites are produced by plants to defend themselves against biotic and abiotic stresses and for the preservation of structure and vivacious functions (Isah, 2019). Due to these reasons plant is considered a new biopesticides. The local and systemic symptoms are found on the plant host (Lucas, 2011). Die-back, root-rot, leaf-spots, damping-off, blight, canker, anthracnose, basal stem rot, soft rot and dry rots and decline are the most common symptoms in plants (Tsitsigiannis et al., 2008). The important fungal phytopathogens populated are given in the Table 26.1. In this chapter a literature survey of the antifungal potential against pathogenic fungi gained from various plant species and used for sustainable tool to management phytopathogens is done.

26.2 SECONDARY METABOLITES OF PLANTS WITH ANTIFUNGAL ACTIVITY

Low-molecular weight compounds secreted from the plants are known as secondary metabolites (SMs) that are not essential for supporting

Table 26.1 Important fungal phytopathogens

Medicinal Plant Species	Plant part	Pathogens	Plant disease	Observations	References
Adenophyllum aurantium, Cleome viscosa	Seeds, leaves	*Alternaria alternata, A. arborescens, A. tenuissima, A. solani, Alternaria* sp., *Alternaria* spp.,	Caused fruit rot of blueberry, apple, pear, melon, apricot, grapes. Spoiled tomatoes. Early blight on potato.	Sporulation and mycelial growth inhibition by methanol and ethyl acetate extracts. Growth and germination inhibition by ethanol extracts.	Moccellin et al., 2018; Abdessemed et al., 2019.
Ageratum conyzoides, Hedeoma multiflora	Leaves	*Aspergillus brasiliensis, A. costaricaensis, A. flavus, A. lacticoffeatus, A. nidulansand A. niger.*	Black and white pepper. Nut kernels. Black rot of grapes, due to ochratoxin A contamination of grapes, grapes by-products, dried vine fruits, coffee, and cocoa.	Essential oils inhibited mycelial growth and aflatoxin production. Germination and growth inhibition by essential oils in combination with those of *Rosmatinus officinalis* L.	Tyśkiewicz et al., 2019; Cucu et al., 2020.
Zingiber officinale	Root	*Bipolaris oryzae and B. sorokiniana*	Caused foliar disease, grain discolouration on cereals in rice, maize and sorghum.	Growth inhibition of aqueous extract and essential oil	Van Bockhaven et al., 2015; Jatoi et al., 2019.

Contd...

Piper holtoni, P. auritum, P. hispidum	Leaves, root, fruit	*Botryodiplodia theobromae*	Infection of cocoa beans	Growth inhibition by n-hexane extract and essential oil.	Żyżelewicz et al., 2019
Amyris texana	Leaves, flower, fruit	*Botrytis cinerea*	Spoiled tomatoes. An important disease of table grapes (*Vitis vinifera L.*)	Growth inhibition by a chromene amide	González-Domínguez et al., 2019
Allium sativum	Fruit, leaves	*Colletotrichum acutatum, C. coccodes, C. fragariae, C. graminicola, C. gloeosporioides, C. lindemuthianum,*	Citrus post-bloom fruit drop epidemics and legume species. Caused anthracnose with black sunken lesion.	Growth, germination and sporulation inhibition of essential oils in vitro and in vivo assays	Bahri et al., 2019
Cupressus benthamii, Vetiveria zizanioides	Leaves, roots	*Phytophthora infestans, P. plurivora and P. ramorum.*	Causal agent of the late blight in solanaceous plant (potato and tomato).	Inhibition of sporangial germination	Okolle et al., 2018
Caryaillinoensis, Ocoteaquixos (Lam.)	Leaves, shells, husk	*Pythium spand P. ultimum*	Variety of diseases including seed rots and damping-off, root, stem and fruit rots, foliar blights.	Growth inhibition by polyphenolic extracts and by essential oils (trans-cinnamaldehyde, methylcinnamate, 1, 8-cineole)	Yaribeygi et al., 2019
Larrea tridentata	Root, leaves	*Rhizoctonia solani, F. vertillioides, F. sambucinum,*	Causes black scurf on potato tubers, stolon and root pruning.	Extracts with polyphenols and lanolin showed growth inhibition.	Guerra et al., 2018.

Chenopodium ambrosiodes, Teloxys ambrosoides	Leaves	*Fusarium avenaceum, F. culmorum, F. graminearum, F. decemcellulare, F. moniliforme,*	Causes of wilts and scab or blight diseases of small grains and ear rot in wheat, barley.	Growth inhibition by essential oils (α-terpinene, p-cymene, 4-carene)	Pintea et al., 2018.
Glycine max, Flourensia cernua	Seeds, leaves, flowers	*Penicillium expansum, P. herquei and Curvularia lunata*	Post-harvest fruit rot disease of apple.	Growth and germination	van Wijck et al., 2018

life but essential as a food material for the microorganisms (Pauli, 2006). The secondary metabolites of plants are synthetized in specific pathway and sites of production can fluctuate between varieties of complexes as well as between the plant species (Grzesiak et al., 2013). In resin ducts, laticifers, oil cells, trichomes, or in cuticles the lipophilic SMs are normally sequestered and the hydrophilic compounds are mainly stored in the vacuole (Wink, 2006). According to their biosynthetic origin in plants the secondary metabolites can be distributed into three main categories are known as: terpenoids, nitrogen-containing compounds (alkaloids, glucosinolates and cyanohydrins) and phenylpropanoids (Springob & Kutchan, 2009). In the biosynthesis of SM shikimic acid, mevalonic acid, acetyl coenzyme A, and 1-deoxyxylulose-5-phosphate are the derivatives and consumed respectively in the acetate, mevalonate, shikimate, and deoxyxylulose phosphate pathways, and also play an important role in chemical signalling and in plant defence mechanisms (Lichtenthaler, 2010). A derivative of plants with antifungal activity the secondary metabolites may performed as inhibitors that are contemporary constitutively in healthy plants or stress circumstances (phytoalexins) (Ansar et al., 2017). The phytopathogen management is discussed in Table 26.2 secreted by plant secondary metabolites.

26.3 INDUCED LOCAL DEFENCE BY MEANS OF PHYTOALEXINS

A group of structurally diverse molecules, known as phytoalexins, are lipophilic and non-specific in their antifungal activities. Once the pathogens are detected in the plant the synthesis of phytoalexins involves transcriptional and/or translational activity (Zhao et al., 2005). The trafficking and secretion of these compounds represented induced response mechanism to the infection site. The compounds that require the novo expression of genes elaborated in the metabolic pathway usually restricted by the phytoalexins (Kwon et al., 2008). The phytoalexins synthesis completed only in early period of infection and localized only in certain area. This is a cost-effective way to neutralize the fungal pathogens. Though, the production of phytoalexins responds to pathogen attack for both disease-resistant and disease-susceptible plants and in the resistant plants these compounds generally accumulate fast at high levels (Lyon et al., 1995).

Table 26.2 Management of crop pathogens by plant secondary metabolites

Classification of phytoalexins	Types of phytoalexins	Example of phytoalexins	Plant species	Managed crop pathogens	References
Terpenes	Monoterpene, Sesquiterpenes, Diterpenes, Sesterpenes, Triterpenes, Sesquarterpenes,	Geraniol, Humulene, Cafestol, Geranylfarsol, Squalene, Ferrugicadiol,	*Saturejathymbra, Thymbraspicata, Helianthus* spp.	*Rhyzoctonia solani, Fusarium moniliforme, Sclerotinium sclerotiorum, Phytophthora capsica, Botrytis cinerea*	Karpiński et al., 2020.
Phenolics	Coumarin, Furano-coumarins, Lignin, Flavonoids, Isoflavonoids,	Hydroxycoumarins, Psoralin, Resveratrol, Quercitin, Genistein,	*Sorghum bicolor, Phaseolus vulgaris, Glicine max, Pisumsativum, Medicago sativa*	*Colletotrichum graminicola, Colletotrichum falcatum, Colletotrichum lindemuthianum, Fufarium* spp., *Aternaria* spp.	Li et al., 2015.
Nitrogen-containing compounds	Alkaloids, Cyanogenic glucosides, Non-protein amino acids	Cocaine, Dhurrin, Canavanin	*Papaver bracteatum, Erucastrum gallicum, Thlaspi arvense*	*Alternaria brassicae, Alternaria alternata, Arabidopsis thaliana, Botrytis species*	Schmidt & Audörsch, 2017.
Sulphur-containing compounds	Glutathiione, Glucosinolate, Thionins, Defensins	B-D-Glucopyrinose	*Brassica rapa, Thlaspi arvense*	*Leptosphaeria maculans, Rhizoctonia solani, Sclerotinia sclerotiorum*	Robin et al., 2017.
Poliketides	Wyerone acid, furanoacetylenic	Acetate units	*Vicia faba*	*Botrytis cinereal, Botrytis fabae*	Bendahmane et al., 2012.

But the less fungitoxic phytoalexins also can accumulate in huge amounts within the plant tissues, beyond the concentrations desired to inhibit the growth of fungi (Roshchina & Roshchina, 1993).

26.3.1 Phenols Phytoalexins

The phenolic acids, coumarins, stilbenes, flavonoids, isoflavonoids, quinones and xanthones compounds are produced by the modifications of aromatic acids and phenolic derived from the metabolites (Crozier & Jaganath, 2009). The *Phaseolus vulgaris, Glicine max, Pisumsativum,* and *Medicago sativa* have been categorized as isoflavonoids phytoalexins. Maackiain from chickpea and pisatin produce from pea plants (Aoki et al., 2000). Alfalfa and soybean plant produced isoflavone daidzein the precursor of major phytoalexins including medicarpin and glyceollins. The phaseollin production is higher in susceptible plants which were resistant to *Colletotrichum lindemuthianum* fungus, the causal agent of anthrachnose disease (Blount et al., 1992). An antraquinone known as Emodin was extremely effective against spore germination of 17 confirmed fungi species, with three species of *Fusarium* and seven species of *Alternaria* and was also extremely inhibitory against a pathogenic basidiomycete fungus *Fomes annosus* isolated from *Rhammus triquetra* bark (Molnar et al., 2010). Phytoalexins induced in *Malva sylvestris* L. are Malvone A known as (2-methyl-3-methoxy-5, 6-dihydroxy-1, 4-naphthoquinone) showing the antifungal activity against the plant pathogen *V. dahlia* (Veshkurova et al., 2006). The scoparone (6, 7-dimethoxycoumarin) is the key phytoalexin which influenced defense mechanism of citrus against pathogens like *Phytophthora parasitica, Guignardia citricarpa, Penicillium digitatum* and *Diaporthe citri.* In response to both pathogenic (*Alternaria helianthi*) and non-pathogenic (*Helminthosporium carbonum*) fungi scopoletin and ayapin collected in separated leaves, leaf disks, stems and hypocotyls of sunflower (*Helianthus annuus*) plants (Afek & Sztejnberg, 1988). In leaves and fruits of grapevines synthesized are resveratrol (trans-3, 5, 4¢-trihydroxystilbene) showed antifungal activity against *Rhizopus stolonifer, Plasmopara viticola,* and *Botrytis cinerea* fungal pathogen (Morrissey & Osbourn, 1999).

26.3.2 Terpenoids Act as Phytoalexins

Monoterpenes carvacrol and thymol are synthesized from *Saturejathymbra* and *Thymbraspicata.* Monoterpenes carvacrol and thymol confirmed the strong antifungal activity against the *B. cinerea* (Abbaszadeh et al., 2014), *Colletotrichum acutatum* and *Penicillium italicum.* The major accumulations of phytoalexins in potato tubers are sesquiterpenes, rishitin, lubimin and solavetivone. Sesquiterpene lactones, niveusin B, ethoxyniveusin B and, leptocarpin revealed in *Helianthus* sp., are active fungicidal compounds. Diterpenoid compounds acting as phytoalexins isolated from rice (*Oryza sativa*) exhibited antifungal activity against *M. grisea* fungus is the causal agent of rice blast disease and also infects wheat, barley, and millet crops. The momilactones exhibit antifungal activity against *M. grisea* exhibited by momilactones and appears in the leaves of rice after infection (Peters, 2006; Chandra & Singh, 2016). The diterpenoid phytoalexins are oryzalexins isolated from rice and, mansonones A-F synthesized in sapwood of *Ulmus, American.*

26.3.3 Nitrogen and Sulphur Containing Compounds Phytoalexins

Alkaloids are the nitrogen containing products found in 20% of higher plants. As a phytoalexins few alkaloids have been identified; sanguinarine is the one exception obtained from *Papaver bracteatum* cell suspension cultures used for fungal elicitation and hormonal deprivation both (Croteau et al., 2000). The crucifers interaction with their fungal pathogens brassinin is of great interest in the biosynthetic pathway and its intermediary such as cyclo brassinin, brassilexin and brassicanal A. Wasalexins A and B isolated from *Wasabia japonica,* produced in response to the blackleg fungus. *Leptosphaeria maculans* is another example of crucifer phytoalexins (Pedras et al., 2007; 2011). Caulilexins A, B, and C are the phytoalexins synthesized from *B. oleracea var. botrytis.* The molecule 3-thiazol-2′-yl-indole refers to camalexin isolated from the leaves of the crucifer *Camelina sativa* infected with *Alternaria brassicae* (Eckadt, 2011).

26.3.4 Poliketides as Phytoalexins

Poliketides derived from metabolites of acetate units is phytoalexins. A post-infection defense response produced from *Vicia faba* tissues against fungal pathogens. Wyerone acid and wyerone furanoacetylenic phytoalexin is the low-molecular weight secondary metabolites (Pusztahelyi et al., 2017). In *B. cinerea* lesions the accumulation completed of wyerone acid whereas the phytoalexin starts to accumulate in *Botrytis fabae* lesions but after some time it decreases. *B. fabae* have capability to colonize broad bean tissues looks to be related to its ability to purify broad bean phytoalexins and to reduce their toxic effects (Nawar & Kuti, 2003).

26.4 PLANT BASED NATURAL PRODUCT AGAINST PLANT PATHOGENS

As an alternative form of healthcare the search for active compounds in plants is very important. To choosing sources of natural products there are several approaches for the potential antifungal compounds detection (Altemimi et al., 2017). To investigate whole extracts of potential antifungal plants and to obtain biological material for fractionation testing are two different approaches. (Svetaz et al., 2010).

26.4.1 Plant Extracts

The common scabs, caused in potato due to the *Streptomyces scabies*, inhibited the growth by *Geranium pratense*, .*Reynoutria suchalinensis* exhibited favourable protecting control of powdery mildew in aqueous extract (Sarwar et al., 2018). The growth of *Aspergillus niger*, *Penicillium ochrochloron*, *Trichoderma viride*, *Fusarium tricinctum* and *Phomopsis helianthi* inhibited by an ethanol extract of *Phlomis fruticosa*. At a very low concentration this extract also showed the fungicidal activity against *Cladosporium cladosporioides* and *Aspergillus ochraceus* (Soković et al., 2013). A dealcoholized extract of *Cassia tora* leaves, obtained against *Aspergillus niger*, showed strong antifungal activity (Phongpaichit et al., 2004). The extracts of *Acer nikoense*, *Glycyrrhiza glabra* and *Thea sinensis* were the most effective plants in very low amounts against antifungal activity. The antimicrobial activity was assessed using a microdilution technique against *Aspergillus niger*, *A. ochraceus*, *A. flavus*, *Penicillium ochrochloron*, *P. funiculosum*, *Trichoderma viride*, *Fusarium tricinctum* and *Alternaria alternate* by the cardoon extracts (Omidpanah et al., 2015). *Bryum argenteum* secreted ethanol which showed antifungal activity against *A. niger* and *P. ochrochloron* fungi (Deora & Guhil, 2014).

26.4.2 Essential Oils

The extracted essential oil of *Origanum majorana* and *O. syriacum* exhibited antifungal activity against the *Aspergillus flavus*, *A. niger*, *A. ochraceus*, *A. parasitcus* and *Trichoderma viride*, *Penicillium* sp., and *Fusarium* sp. (Deans & Svoboda, 1990). Allspice and cloves essential oils inhibited totally the growth of *Trichoderma viride*, *Alternaria alternata*, *Fusarium oxysporum*, *Mucor circinelloides*, *Rhizopus stolonifer*, *Cladosporium cladosporioides*, *Aspergillus versicolor* and *Penicillium citrinum* fungal species using with 2% concentration (Wei et al., 2013; Chandra & Enespa, 2017). More effective antifungal activity was shown by the essential oil of *Piper angustifolium* (Kalemba & Kunicka, 2003). The oil of *Monarda citriodora* and *Melaleuca alternifolia* exhibited a high level of antifungal activity directly contact and in vapour form against rapidly growing fungal species (Al Yousef, 2014). The antifungal activity testing against phytopathogenic fungi *C. cladopsorioides* and *P. helianthi* more effective due to the essential oil of *Foeniculum vulgare* (Mimica-Dukić et al., 2003). The essential oil of *Cedrus deodara* and *Trachysremum ammi* showed fungitoxicity against *Aspergillus niger* and *Curvularia ovoides* and can be used as fungicide (Buneri et al., 2019). The growth of *C. capsici* fungus inhibited by cumaldehyde and eugenol oil against the Rips tot and die Back diseases. Cineole, and eugenol oils inhibited the growth of *P. psidi and R. nadssus. F. monoliformae* fungi caused against "softrot" of apical tissue, head blight, scab and root rot of wheat (Stojković et al., 2011; Vet, 2003).

26.5 CONCLUSION

The naturally produced fungicides are considered more eco-friendly than the man-made herbicides, because of their short environmental half-life. The phytochemicals of plants have inhibitory effects on all types of microorganisms in vitro. Approximately 600 plants showing antifungal properties have been reported till date. The extraction of phytochemicals profile may vary on the bases of climatic, geographical condition or genetic factors. Finally, the sustainable food security cannot rely on the use of synthetic pesticides; endemic resources of biopesticides must be utilized in the production and post-harvest protection of food products in developing countries, in order to ensure that the populations will have access to sufficient and healthy food.

REFERENCES

Abbaszadeh, S.; Sharifzadeh, A.; Shokri, H.; Khosravi, A.R.; Abbaszadeh, A. Antifungal efficacy of thymol, carvacrol, eugenol and menthol as alternative agents to control the growth of food-relevant fungi. *J. Mycol Medical*, 2014, 24(2): e51-e56.

Abdessemed, N.; Kerroum, A.; Bahet, Y.A.; Talbi, N.; Zermane, N. First report of *Alternaria* leaf spot caused by *Alternaria alternata* (Fries.) Kiessler on *Sonchus oleraceus* L. and *Convolvulus arvensis* L. in Algeria. *J. Phytopathol*, 2019, 167(6): 321-325.

Afek, U.; Sztejnberg, A. Accumulation of scoparone, a phytoalexin associated with resistance of citrus to *Phytophthora citrophthora*. *Phytopathol*, 1988, *78*(12): 1678-1682.

Al Yousef, S. A. Essential oils: their antimicrobial activity and potential application against pathogens by gaseous contact a review. *Egypt. J. Biol. Sci, G. Microbiol*, 2014, 6(1): 37-54.

Altemimi, A.; Lakhssassi, N.; Baharlouei, A.; Watson, D.C.; Lightfoot, D.A. Phytochemicals: Extraction, isolation, and identification of bioactive compounds from plant extracts. *Plants*, 2017, 6(4): 42.

Ansar, M.; Ghatak, A.; Ghatak, L.V. Secondary Metabolites in Pathogen-Induced Plant Defense. In *Plant Secondary Metabolites, Volume Three.* Apple Academic Press. 2017; pp. 189-204.

Aoki, T.; Akashi, T.; Ayabe, S.I. Flavonoids of leguminous plants: structure, biological activity, and biosynthesis. *J. Plant Res.*, 2000, 113(4): 475.

Bahri, B. A.; Saadani, M.; Mechichi, G.; Rouissi, W. Genetic diversity of *Colletotrichum gloeosporioides* species complex associated with *Citrus* wither-tip of twigs in *Tunisia* using microsatellite markers. *J. Phytopathol.*, 2019, 167(6): 351-362.

Bendahmane, B.S.; Mahiout, D.; Benzohra, I.E.; Benkada, M.Y. Antagonism of three *Trichoderma* species against *Botrytis fabae* and *B. cinerea*, the causal agents of chocolate spot of faba bean (*Vicia faba* L.) in Algeria. *World Appl. Sci. J.*, 2012, 17(3): 278-283.

Blount, J.W.; Dixon, R.A.; Paiva, N.L. Stress responses in alfalfa (*Medicago sativa* L.) XVI. Antifungal activity of medicarpin and its biosynthetic precursors; implications for the genetic manipulation of stress metabolites. *Physiol. Mol. Plant Pathol.*, 1992, 41(5): 333-349.

Buneri, I.D.; Yousuf, M.; Attaullah, M.; Afridi, S.; Anjum, S.I.; Rana, H.; Ansari, M.J.A comparative toxic effect of *Cedrus deodara* oil on larval protein contents and its behavioral effect on larvae of mealworm beetle (*Tenebrio molitor*)(*Coleoptera: Tenebrionidae*). *Saudi J. Biol. Sci.*, 2019, 26(2): 281-285.

Chandra, P.; Enespa. Microbial volatiles as chemical weapons against pathogenic fungi. In: *Volatiles and Food Security*, 2017, (pp. 227-254). Springer, Singapore.

Chandra P.; Singh, E. Applications and mechanisms of plant growth-stimulating rhizobacteria. In: *Plant-Microbe Interaction: An Approach to Sustainable Agriculture*, 2016, (pp. 37-62). Springer, Singapore.

Chithra, S.V.; Nair, M.H.; Amarnath, A.; Anjana, N.S. Impacts of impervious surfaces on the environment. *Int. J. Eng. Sci. Invention*, 2015, 4(5): 2319-6726.

Croteau, R.; Kutchan, T.M.; Lewis, N.G. Natural products (secondary metabolites). *Biochem. Mol. Biol. Plant*, 2000, 24: 1250-1319.

Crozier, A.; Jaganath, I.B.; Clifford, M.N. Dietary phenolics: chemistry, bioavailability and effects on health. *Nat. Prod. Rep.*, 2009, 26(8), 1001-1043.

Cucu, M.A.; Gilardi, G.; Pugliese, M.; Gullino, M.L.; Garibaldi, A. An assessment of the modulation of the population dynamics of pathogenic *Fusarium oxysporum* f. sp. lycopersici in the tomato rhizosphere by means of the application of *Bacillus subtilis* QST 713, *Trichoderma* sp. TW2 and two composts. *Biol. Cont.*, 2020, 142: 104158.

Dayan, F.E.; Duke, S.O. Natural compounds as next-generation herbicides. *Plant Physiol.*, 2014, 166(3): 1090-1105.

Deans, S.G.; Svoboda, K.P. The antimicrobial properties of marjoram (*Origanum majorana* L.) volatile oil. *Flavour Frag J.*, 1990, 5(3): 187-190.

Deora, G.S.; Guhil, N. Bryophytes: A potent tool for controlling some fungal diseases of crop plants. *Int. J. Pharma. Sci. Invention*, 2014, 3(6): 21-26.

Eckadt, N.A. Induction of Phytoalexin Biosynthesis: WRKY33-Is a Target of MAPK Singling. *Plant Cell*, 2011, 23(4): 1190.

Engelmeier, D.; Hadacek, F. Antifungal natural products: assays and applications. *Adv. Phytomed.*, 2006, 3: 423-467.

García-Mier, L.; Guevara-González, R.G.; Mondragón-Olguín, V.M.; del Rocío Verduzco-Cuellar, B.; Torres-Pacheco, I. Agriculture and bioactives: achieving both crop yield and phytochemicals. *Int J. Mol. Sci.*, 2013, 14(2): 4203-4222.

González-Domínguez, E.; Fedele, G.; Languasco, L.; Rossi, V. Interactions among fungicides applied at different timings for the control of *Botrytis* bunch rot in grapevine. *Crop Protection*, 2019, 120: 30-33.

Grzesiak, M.; Filek, M.; Barbasz, A.; Kreczmer, B.; Hartikainen, H. Relationships between polyamines, ethylene, osmoprotectants and antioxidant enzymes activities in wheat seedlings after short-term PEG-and NaCl-induced stresses. *Plant Growth Regul.*, 2013, 69(2): 177-189.

Guerra, A.R.; Duarte, M.F.; Duarte, I.F. Targeting tumor metabolism with plant-derived natural products: Emerging trends in cancer therapy. *J. Agricul. Food Chem.*, 2018, 66(41): 10663-10685.

Isah, T. Stress and defense responses in plant secondary metabolites production. *Biol. Res.*, 2019, 52(1): 39.

Jatoi, G.H.; Keerio, A.U.; Abdulle, Y.A.; Qiu, D. Effect of selected fungicides and Bio-Pesticides on the mycelial colony growth of the *Helminthosporium oryzae*. Brown spot of rice. *Acta. Ecologica. Sinica.*, 2019, 39(6): 456-460.

Kalemba, D.A. A.K.; Kunicka, A. Antibacterial and antifungal properties of essential oils. *Current Med. Chem.*, 2003, 10(10): 813-829.

Karpiński, T. M. Essential Oils of Lamiaceae Family Plants as Antifungals. *Biomol.*, 2020, 10(1): 103.

Khater, H. F. Prospects of botanical biopesticides in insect pest management. *Pharmacologia*, 2012, 3(12): 641-656.

Kwon, C., Bednarek, P.; Schulze-Lefert, P. Secretory pathways in plant immune responses. *Plant Physiol.*, 2008, 147(4): 1575-1583.

Lattanzio, V.; Kroon, P. A.; Quideau, S.; Treutter, D. (2008). Plant phenolics—secondary metabolites with diverse functions. *Recent Adv. Polyphenol Res.*, 2008, 1: 1-35.

Li, H.; Ban, Z.; Qin, H.; Ma, L.; King, A.J.; Wang, G. A heteromeric membrane-bound prenyl transferase complex from hop catalyzes three sequential aromatic prenylations in the bitter acid pathway. *Plant Physiol.*, 2015, 167(3): 650-659.

Lichtenthaler, H. K. Biosynthesis and emission of isoprene, methyl butanol and other volatile plant isoprenoids. "*The Chemistry and Biology of Volatiles*", John Wiley and Sons, 2010, pp. 11-47.

Lucas, J. A. Advances in plant disease and pest management. *J. Agricultural Sci.*, 2011, 149 (S1): 91-114.

Lyon, G.D.; Reglinski, T.; Newton, A.C. Novel disease control compounds: the potential to 'immunize'plants against infection. *Plant Pathol.*, 1995, 44(3): 407-427.

Mimica-Dukić, N.; Kujundžić, S.; Soković, M.; Couladis, M. Essential oil composition and antifungal activity of *Foeniculum vulgare* Mill. Obtained by different distillation conditions. *Phytotherapy Research: J. Devot. Pharmacol. Toxicol. Eval. Nat. Prod. Derivative,* 2003, 17(4): 368-371.

Moccellin, R.; Bellé, C.; Kaspary, T.E.; Groth, M.Z.; Rohrig, B.; Cutti, L.; Casarotto, G. Detection of *Alternaria alternata* Causing Leaf Spot on *Rumex obtusifolius* in Southern Brazil. *Plant Dis.*, 2018, 102(8): 1656-1656.

Molnar, I.; Gibson, D.M.; Krasnoff, S.B. Secondary metabolites from entomopathogenic *Hypocrealean* fungi. *Nat. Prod. Rep.*, 2010, 27(9): 1241-1275.

Morrissey, J.P.; Osbourn, A.E. Fungal resistance to plant antibiotics as a mechanism of pathogenesis. *Microbiol. Mol. Biol. Rev.*, 1999, 63(3): 708-724.

Nawar, H.F.; Kuti, J.O. Wyerone acid phytoalexin synthesis and peroxidase activity as markers for resistance of broad beans to chocolate spot disease. *J. Phytopathol*, 2003, 151(10): 564-570.

Okolle, J.N.; Lum, N.M.C.; Ngosong, C.; Nanganoa, L.T.; Clovis, T.; Peter, N. Efficacy of four plant emulsions for managing long tail mealy bugs (*Pseudococcus longispinus*) infesting banana plantations in Cameroon. *J. Entomol. Zool. Stud.*, 2018, 6(3): 837-843.

Omidpanah, S.; Sadeghi, H.; Sarcheshmeh, M.M.; Manayi, A. Evaluation of antifungal activity of aqueous extracts of some medicinal plants against *Aspergillus flavus,* pistachio aflatoxin producing fungus in vitro. *Drug Devel. Therapeutics*, 2015, 6(2): 66.

Pauli, A. Anticandidal low molecular compounds from higher plants with special reference to compounds from essential oils. *Med. Res. Rev.*, 2006, 26(2): 223-268.

Paulitz, T. C.; Bélanger, R. R. Biological control in greenhouse systems. *Ann. Rev. Phytopathol.*, 2001, 39(1): 103-133.

Pedras, M.S.C.; Yaya, E.E.; Glawischnig, E. The phytoalexins from cultivated and wild crucifers: chemistry and biology. *Natural Prod. Report.*, 2011, 28(8): 1381-1405.

Pedras, M.S.C.; Zheng, Q.A.; Sarma-Mamillapalle, V.K. The phytoalexins from Brassicaceae: structure, biological activity, synthesis and biosynthesis. *Natural Product Communi*, 2007, 2(3): 1934578X0700200315.

Peters, R.J. Uncovering the complex metabolic network underlying diterpenoid phytoalexin biosynthesis in rice and other cereal crop plants. *Phytochem.*, 2006, 67(21): 2307-2317.

Phongpaichit, S.; Pujenjob, N.; Rukachaisirikul, V.; Ongsakul, M. Antifungal activity from leaf extracts of *Cassia alata* L., *Cassia fistula* L. and *Cassia tora* L. *Songklanakarin J. Sci. Technol.*, 2004, 26(5): 741-748.

Pintea, A.; Rugină, D.; Diaconeasa, Z. Pharmacologically Active Plant-Derived Natural Products. *Nanoparticles Biomed.*, Elsevier. 2018 (pp. 49-64).

Pusztahelyi, T.; Holb, I.J.; Pócsi, I. Plant-fungal interactions: special secondary metabolites of the biotrophic, necrotrophic, and other specific interactions. *Fungal Metabol.*, 2017, pp. 133-190. Springer, Chapter.

Ribera, A.E.; Zuñiga, G. Induced plant secondary metabolites for phytopatogenic fungi control: a review. *J. Soil Sci. Plant Nutrition*, 2012, 12(4): 893-911.

Robin, A.H.K.; Yi, G.E.; Laila, R.; Hossain, M.R.; Park, J.I.; Kim, H.R.; Nou, I.S. *Leptosphaeria maculans* alters glucosinolate profiles in blackleg disease–resistant and-susceptible cabbage lines. *Frontiers in Plant Science*, 2017, 8: 1769.

Roshchina, V.V.; Roshchina, V.D. The elimination of substances in response to extreme factors. In: *The Excretory Function of Higher Plants*. Springer, Berlin, Heidelberg. 1993, (pp. 177-207).

Saini, R.K.; Patel, S.; Bajpai, J.; Bajpai, A.K. Advanced controlled nanopesticide delivery systems for managing insect pests. In: *Controlled Release of Pesticides for Sustainable Agriculture.*, Springer, Cham. 2020, (pp. 155-184).

Sarwar, A.; Latif, Z.; Zhang, S.; Zhu, J.; Zechel, D.L.; Bechthold, A. Biological control of potato common scab with rare isatropolone C compound produced by plant growth promoting *Streptomyces* A1RT. *Front. Microbiol.*, 2018, 9: 1126.

Schmidt, B.; Audörsch, S. Stereoselective Total Syntheses of Polyacetylene Plant Metabolites via Ester-Tethered Ring Closing Metathesis. *J. Organic Chem.*, 2017, 82(3): 1743-1760.

Soković, M.D.; Glamočlija, J.M.; Ćirić, A.D. Natural products from plants and fungi as fungicides. *Fungi. Showcases Integrated Plant Disease Manag. World*, 2013, 14: 185-232.

Springob, K.; Kutchan, T.M. Introduction to the different classes of natural products. *Plant-Derived Natural Prod.* Springer, New York, NY. 2009, (pp. 3-50).

Stojković, D.; Soković, M.; Glamočlija, J.; Džamić, A.; Ćirić, A.; Ristić, M.; Grubišić, D. Chemical composition and antimicrobial activity of *Vitexagnus-castus* L. fruits and leaves essential oils. *Food Chem.*, 2011, 128(4): 1017-1022.

Svetaz, L.; Zuljan, F.; Derita, M.; Petenatti, E.; Tamayo, G.; Cáceres, A.; Gupta, M. Value of the ethnomedical information for the discovery of plants with antifungal properties. A survey among seven Latin American countries. *J. Ethnopharmacol.*, 2010, 127(1): 137-158.

Tsitsigiannis, D.I.; Antoniou, P.P.; Tjamos, S.E.; Paplomatas, E.J. Major diseases of tomato, pepper and egg plant in green houses. *Europ. J. Plant Sci. Biotechnol*, 2008, 2(1): 106-124.

Tyśkiewicz, K.; Tyśkiewicz, R.; Konkol, M.; Rój, E.; Jaroszuk-Ściseł, J.; Skalicka-Woźniak, K. Antifungal Properties of *Fucusvesi culosus* L. Supercritical Fluid Extract Against *Fusarium culmorum* and *Fusarium oxysporum*. *Molecules*, 2019, 24(19): 3518.

Van Bockhaven, J.; Spíchal, L.; Novák, O.; Strnad, M.; Asano, T.; Kikuchi, S.; De Vleesschauwer, D. Silicon induces resistance to the brown spot fungus *Cochliobolus miyabeanus* by preventing the pathogen from hijacking the rice ethylene pathway. *New Phytologist*, 2015, 206(2): 761-773.

vanWijck, Y.; Hiemstra, P.S.; Taube, C. Response to comment on "Therapeutic application of an extract of *Helicobacter pylori* ameliorates the development of allergic airway disease". *J. Immunol.*, 2018, 200(9): 3027-3028.

Veljić, M.; Đurić, A.; Soković, M.; Ćirić, A.; Glamočlija, J.; Marin, P.D. Antimikrobna aktivnost metanolnih ekstrakata *Fontinalis antipyretica, Hypnum cupressiforme Ctenidium molluscum*. *Archi. Biol. Sci.*, 2009, 61(2): 225-229.

Veshkurova, O.; Golubenko, Z.; Pshenichnov, E.; Arzanova, I.; Uzbekov, V.; Sultanova, E., Stipanovic, R.D. Malvone A, a phytoalexin found in *Malva sylvestris*. *Phytochem.*, 2006, 67(21): 2376-2379.

Vet, J. Pesticidal effect of Akee Apple against stored-product insect pests. *Vet J*, 2003, 80: 848-850.

Vurro, M.; Bonciani, B.; Vannacci, G. Emerging infectious diseases of crop plants in developing countries: impact on agriculture and socio-economic consequences. *Food Security*, 2010, 2(2): 113-132.

Wei, D.; Schmidt, O.; Liese, W. Durability test of bamboo against fungi according to EN standards. *Europ. J. Wood Wood Product*, 2013, 71(5): 551-556.

Wink, M. Importance of plant secondary metabolites for protection against insects and microbial infections. *Adv. Phytomed*, 2006, 3: 251-268.

Yaribeygi, H.; Simental-Mendía, L.E.; Butler, A.E.; Sahebkar, A. Protective effects of plant-derived natural products on renal complications. *J. Cell. Physiol.* 2019, 234(8): 12161-12172.

Zhao, J.; Davis, L.C.; Verpoorte, R. Elicitor signal transduction leading to production of plant secondary metabolites. *Biotechnol Adv.*, 2005, 23(4): 283-333.

Żyżelewicz, D.; Bojczuk, M.; Budryn, G.; Jurgoński, A.; Zduńczyk, Z.; Juśkiewicz, J.; Oracz, J. Influence of diet enriched with cocoa bean extracts on physiological indices of laboratory rats. *Molecules*, 2019, 24(5): 825.

27 CHAPTER

Scope of Botanical Pesticides in Storage Insect Pests Management

Sidra Sarwar[1] and Muhammad Sarwar[2*]
[1] Department of Botany, Government Post Graduate College, Faisalabad, Pakistan
[2*] National Institute for Biotechnology & Genetic Engineering (NIBGE), Faisalabad, Pakistan.
Corresponding author: E-mail: drmsarwar64@yahoo.com

ABSTRACT

Losses of food grains due to insects infestation during storage is a serious problem, particularly in the developing countries. In general, the major pests of stored grain are classified into primary pests, which are capable of penetrating and infesting intact kernel of grain and have immature stages development within kernel of grain, and secondary pests, which cannot infest the whole grain but feed on broken kernels and grain damaged by primary pests and their immature stages are found external to the grain. A high percent infestation of grains due to various insect pests during storage causes loss of germination capacity of seeds making them unfit for human consumption. Insect induced changes in the storage environment may cause warm, moist 'hotspots' that are suitable for the development of storage fungi which cause further losses. The use of synthetic chemical insecticides is either not permitted or these are used restrictively because of the residue problem and health risks to consumers. Such constraints pave a way to use need based plant derived products as alternatives to current chemical pesticides. A high volume of reports has shown plant-derived products to possess great insecticidal potential. Moreover, the biosafety studies of various plant based insecticides should be carried out to ascertain their toxic effects on humans, animals and agricultural crops. This chapter includes various aspects like storage losses by insect pests, plant products as alternative to synthetic compounds, and classification of plant based compounds, powder and oil formulations as grain protectants.

27.1 INTRODUCTION

Internationally, it has been estimated that each year between one quarter to one-third of the world grain crops are lost during storage. Losses of grain in storage due to insect are the final components of the struggle to limit insect losses in agricultural production and these losses can exceed those incurred while growing the crop. Worldwide, losses in stored products, caused by insects, have been estimated to be between five and ten per cent and heavier losses occurring in the tropics may reach 30% annually. Losses caused by insects include not only the direct consumption of kernels, but also include accumulations of frass, exuviae, webbing and insect cadavers. In addition, grain which is not lost is severely reduced in quality by insect damage. Many grain pests preferentially eat out grain embryos, thereby reducing the protein content of food grain and lowering the percentage of seeds germination. High levels of this insect detritus may result in grain that is unfit for human consumption. Insect induced changes in the storage environment may cause warm, moist 'hotspots' that are suitable for the development of storage fungi which cause further losses (Ali et al., 2011; Sarwar, 2015a; 2016a).

Economically speaking, storage insects and to a lesser degree, fungi reduce the quality and value of grain, while losses due to rodents and birds are typically quite infrequent and minor (Sarwar, 2016b).

Some important stored grain pests include the lesser grain borer, rice weevil and rust red flour beetle. In fact, grain insect pests may be divided into primary grain insects [rice weevil *Sitophilus oryzae* (L.) (Fig. 27.1), granary weevil *Sitophilus granaries* (L.), (Coleoptera: Curculionidae (Fig. 27.2), lesser grain borer *Rhyzopertha dominica* (F.), (Coleoptera: Bostrichidae Fig. 27.3), Khapra beetle *Trogoderma granarium* (Everts), (Coleoptera: Dermestidae Fig. 27.4), and pulse beetle *Callosobruchus chinensis* (L.), (Coleoptera: Bruchidae Fig. 27.5)], have the ability to attack whole and unbroken grains, and their immature stages development is within kernel of grain (Sarwar, 2015b; 2015c).

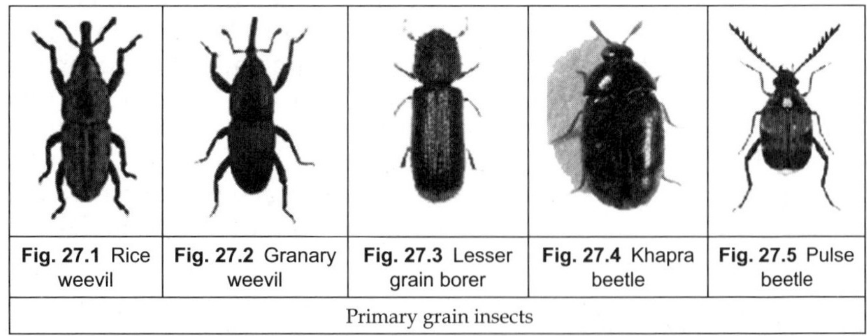

Fig. 27.1 Rice weevil | **Fig. 27.2** Granary weevil | **Fig. 27.3** Lesser grain borer | **Fig. 27.4** Khapra beetle | **Fig. 27.5** Pulse beetle

Primary grain insects

The secondary pests [rust-red flour beetle *Tribolium castaneum* (Herbst), (Coleoptera: Tenebrionidae Fig. 27.6), rusty grain beetle *Cryptolestes ferrugineus* (L.), (Coleoptera: Cucujidae Fig. 27.7), sawtoothed grain beetle *Oryzaephilus surinamensis* (L.), (Coleoptera:

Silvanidae Fig. 27.8), grain mite (Acarina) and psocid *Liposcelis corrodens* (Psocoptera: Liposcelidae Fig. 27.10)] attack only damaged grain, dust and milled products. Generally, the immature stages of the secondary pest species are found external to the grain (Deeba et al., 2006; Ali et al., 2009).

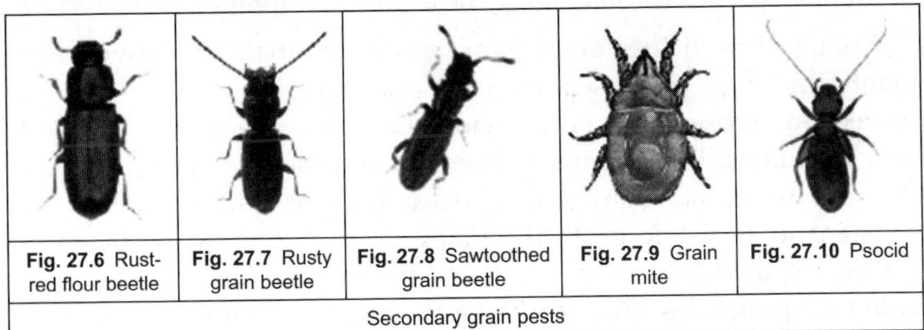

Fig. 27.6 Rust-red flour beetle | **Fig. 27.7** Rusty grain beetle | **Fig. 27.8** Sawtoothed grain beetle | **Fig. 27.9** Grain mite | **Fig. 27.10** Psocid

Secondary grain pests

The insect species of most importance in the tropics (**hot climates)** as pests of stored foodgrains including pulses (legume grains), such as Coleoptera and Lepidoptera are listed in Table 27.1. Many other insects, for instance **Psocoptera,** may occur quite commonly and sometimes abundantly on stored cereal grains especially when they are underdried condition or have been heavily infested by the major primary and secondary pests (Sarwar, 2009; 2013a)

Table 27.1 Important insect pests of tropical stored grains or grain products

Order	Family	Insect pests
Coleoptera	Anobiidae, Bostrichidae, Bruchidae, Cucujidae, Curculionidae, Dermestidae, Silvanidae, Tenebrionidae	*Lasioderma serricorne* (F.), *Rhyzopertha dominica* (F.), *Prostephanus truncatus* (Horn.), *Acanthoscelides obtectus* (Say), *Callosobruchus* spp., *Zabrotes subfasciatus Boheman, Cryptolestes* spp., *Sitophilus oryzae* (L.), *S. zeamais Motschulsky, Trogoderma granarium Everts, Dermestes* spp., *Oryzuephilus surinamensis* (L.), *Tribolium castaneum* (Herbst)
Lepidoptera	Gelechiidae, Pyralidae	*Sitotroga cerealella* (Olivier), *Ephestia cautella* (Walker), *Plodia interpunctella* (Hubner), *Corcyra cephalonica* (Stainton)
Psocoptera	Liposcelidae	*Liposcelis* sp.

Limiting of insect infestation in grain storage must be a primary consideration at the time of harvest. In several geographical areas of the world, different traditional and indigenous methods are employed to control the insect damage in stored grains.Many plants and minerals have insecticidal properties; that is, they are toxic to insects. Botanical insecticides are naturally occurring chemicals (insect toxins) extracted or derived from plants or minerals. They are also called natural insecticides. Organic gardeners choose these insecticides, in some cases, over synthetic organic materials (Sarwar, 2015d). This chapter will help the producers to storing grain on-farm and the commercial elevator operators to become familiar with the available methods for managing these pests in storage.

Botanical pesticides are organic pesticides that are derived from plants and typically used by farm managers as a form of grain protection. These types of pesticides have become more popular as they don not release toxins as they decompose and have proven to be more useful than conventional insecticides as insects become more resistant to synthetic pesticides.Plants use self-created pesticides as a form of defense against diseases, herbivores, and insects. These botanical pesticides include chemicals such as nicotine and oils like citrus oils. The insecticides made with these natural pesticides have minimal to no toxic effects on the environment while acting fast to tackle insects (Ujvary, 1999).

Several hundred candidate plant species and compounds are now known to have pesticidal properties against a range of arthropod pest species.The botanical insecticides that have primarily been used and are commercially available include ryania (Fig. 27.11. *Ryania speciosa*), rotenone (Fig. 27.12. *Lonchocarpus* **sp**.), pyrethrin (Fig. 27.13. *Tanacetum cinerariaefolium*), nicotine (Fig. 27.14. *Nicotiana tabacum*), azadirachtin (Fig. 27.15. *Azadirachta indica*) and sabadilla (Fig. 16. *Schoenocaulon officinale*) (Table 27.2).

Table 27.2 A list of insecticidal active ingredients of some plant materials.

Plant species	Active ingredient	Insect toxicity	Insect specie
Annona reticulate	Anonaine	Contact	*Callosobruchus chinensis*
Azadirachta indica	Azadirachtin	Contact, insect growth regulator	Stored grain pests, aphids
Foeniculum vulgare	E-Anethole	Contact	*Sitophilus oryzae, Callosobruchus chinensis*
Acorus calamus	β-Asarone	Contact	*Stored grain pests*
Acorus calamus, Acorus gramineus	Z-Asarone	Contact	*Sitophilus zeamais*
Chamaecyparis obtusa	Bornyl acetate	Contact	*Sitophilus oryzae*

Ocimum kilimandscharicum	Camphor	Contact	*Sitophilus oryzae*
Baccharis salicifolia	(+)-3-Carene	Contact	*Tribolium castaneum*
Thujopsis dolabrata	Carvacrol	Contact; fumigant	*Sitophilus oryzae, Callosobruchus chinensis*
Carum carvi	Carvone	Contact	*Sitophilus oryzae, Rhyzopertha dominica*
Eucalyptus	1, 8 Cineole	Contact, fumigant	*Rhyzopertha dominica, Tribolium castaneum*
Cinnamomum aromaticum	Cinnamal-dehyde	Contact	*Tribolium castaneum,Sitophilus zeamais*
Conyza dioscoridis	Diocty lhexanedioate	Contact	*Tribolium castaneum, Sitophilus granaries*
Citrus	Eugenol	Fumigant	*Sitophilus oryzae*
Foeniculum vulgare	Estragole	Contact	*Sitophilus oryzae, Lasioderma serricorne*
Foeniculum vulgare	(+)-Fenchone	Contact	*Sitophilus oryzae, Lasioderma serricorne*
Chenopodium ambrosioides	Hexadecane	Contact	*Tribolium castaneum, Sitophilus granaries*
Convolvulus arvensis	Hexadecanoic acid	Contact	*Sitophilus oryzae, Rhyzopertha dominica*
Ocimum canum Sims	Linalool	Fumigant	*Tribolium castaneum, Sitophilus granaries*
Citrus	Limonene	Contact	*Tribolium castaneum*
Baccharis salicifolia	(−)-Limonene	Contact, fumigant	*Tribolium castaneum*
Nicotiana tabacum	Nicotine	Contact	Mites, aphids, thrips, leafhoppers
Tanacetum cinerariaefolium	Pyrethrin I and II	Contact, stomach poison	Stored grain pests, crop pests
Baccharis salicifolia	β-Pinene	Contact	*Tribolium castaneum*
Baccharis salicifolia	α-Pinene	Fumigant	*Tribolium castaneum*
Lonchocarpus sp.	Rotenone	Contact, stomach poison	Sitophilus oryzae, crop pests, lace bugs
Ryania speciosa	Ryania	Contact, stomach poison	Stored grain pests, potato beetle, aphids, lace bugs

Contd...

Schoenocaulon officinale	Sabadilla	Contact, stomach poison	Caterpillars, stinks, thrips, squash bugs, leaf hoppers
Saccharopolyspora spinosa	Spinosyn A and D	Stomach poison	Stored grain pests

Fig. 27.11 *Ryania speciosa*

Fig. 27.12 *Lonchocarpus* sp.

Fig. 27.13 *Tanacetum cinerariaefolium*

Fig. 27.14 *Nicotianata bacum*

Fig. 27.15 *Azadira chta indica*

Fig. 27.16 *Schoenocaulon officinale*

Some insecticidal plants

27.2 BOTANICALS AND VEGETABLE OILS

Plant derived natural grain products are used as powder formulations or liquid or oil formulations against.

27.2.1 Powder Formulations

Neem based products (from the plant *Azadirachta indica J.*, Meliaceae) are highly effective against *Callosobruchus* species. Different doses of neem leaf powder are effective in reducing grain damage by *C. chinensis* in stored lentil. Neem saw dust showed high toxicity to *C. chinensis* on chickpea. Powders *of Acorus calamus* and neem oil are also effective against *C. maculatus* (Isman, 2006).

Powders of mussambe leaves (*Cleome spinosa*), juazerio stem (*Ziziphus jaazeiro*) and pereiro stems (*Aspidosperma pyrifolium*) gave low adult emergence of *C. maculatus* on cowpea seeds (Melo et al., 2010). Powder extract of *Parthenium hysterophorus* inflorescence, stem and leaf on chickpea seeds produced more than 50% mortality of *C. chinensis* (Tesfu and Emana, 2013). Ashes of *Combretum imberbe* have insecticidal properties and can be effectively used for the management of *C. chodesianus* in cowpea storage for smallholders. The powder of *Piptadeniastrum africanum* is toxic to *C. maculatus*, resulting in 100% mortality in cowpea seeds after four days (Ojo and Ogunleve, 2013).

Powdered seeds of *Litsea citrata* are effective against *C. chinensis* for up to two months of storage. Seed powder of *Vittalaria paradoxa* can be used as an effective biopesticide against *C. maculatus* (Abdullahi and Majeed, 2011). Adults of *C. maculates* are susceptible to root bark powder of *Zanthoxylum zanthoxyloides* (Musa, 2012). The powders of *Chenopodium ambrosioides* and *Piper nigrum* reduced both oviposition and adult emergence of *C. maculatus*. Fruit and seed powder of *Capsicum annum* and *C. frutescens* have potential seed protectant activity against *C. maculatus* in cowpea (Oni and Ileke, 2008).

A study has been conducted to determine the protective effects of unripe banana peel liquid extract on cowpea seeds (*Vigna unguiculata*) against cowpea weevil (*Callosobruchus maculatus*). Petroleum ether solvent has been employed in the extraction process which yielded the extract (23.71% w/w). Four concentrations (10%, 1%, 0.1% and 0.01% w/v) of the liquid extracts have been each mixed with 5 g of cowpea seeds put in plastic vials in triplicate. Each experimental unit has ten pairs of adult *C. maculatus*. The effects of different concentrations of each treatment on insect mortality, weight loss and viability of cowpea seeds have been assessed. The highest concentration (10% w/w) induced the highest parentage mortality of *C. maculatus*. There is a significant difference ($P<0.05$) in parentage mortality of weevils in the treated and untreated cowpea seeds subjected to weevils. Unripe banana peel liquid extract gave appreciably reduced average percentage weight loss when compared to the untreated seeds. There is a significant level of weight loss between treated and untreated cowpea seeds. Seeds treated with the highest concentration (10% w/w) of liquid extract showed the least weight loss and is not viable whilst those treated with 1% w/w, 0.1% w/w and 0.01% w/w of liquid extracts are viable, with control having the least percentage viability. Seeds treated with liquid extract at concentration 0.01% w/w prevented infestations by weevil's larvae, and favoured germination more than seeds which are not treated with liquid extract and also not subjected to weevils but kept for the same duration as the experiments (control). Unripe banana peel liquid extract protected cowpea seeds from infestations by the weevil's larvae. Hence, this research supports the local use of unripe banana peel liquid as a protectant and also reveals the concentration of liquid extract which favours germination of cowpea seeds (Boakye et al., 2016).

27.2.2 Liquid or Oil Formulations

Solvent extracts of *Ocimum basilicum* and *Hyptis spicigera* showed pesticidal effect on *C. maculatus*. *Ocimum gratissimum* also possess significant biopesticidal activity against *C. maculatus*. Extracts of *Citrus vsinensis* and *Euphorbia balsamifera* potentially reduced *C. maculatus* infestation in cowpea seeds (Suleiman et al., 2014). Ovicidal activity against *C. chinensis* was reported for foliar extract of *Delonix regia*. The plant latex of *Astonia boonei, Calotropis procera* and *Jatropha curcas* are effective against *C. maculatus* (Ileke et al., 2012).

Essential oils from *Eucalyptus camaldulensis* and *Heradeum persicum* are found to be effective against *C. maculatus*. Oil from *Clausenapen taphylla* is found to be the most effective herbal insecticide against *Calloso bruchus* species. Essential oil of *Menthapu legium* is effective against different stages of *C. maculatus* and essential oil from *Lantana camera* showed strong repellent activity against adults of *C. maculatus, while* leaf powder of *J. curcas* is effective as well (Kenneth et al., 2018). The insecticidal property of essential oil from *Artemisia judaica*, which was effective against *C. maculatus*, was due to the presence of piperitone (32.4%), camphor (20.6%) and (E) ethyl cinnamate (8.2%) (Isman, 2002).

The essential oil of *Artemisia vulgaris* showed fumigant toxicity to *C. maculatus*. Similarly, the fumigant toxicity of essential oil of *Rosmarinus officinalis* significantly affected adult longevity, fecundity and fertility of *C. maculates* (Sharifian et al., 2013).When grain pulses are stored at farms application of vegetable oil can give good protection against *C. maculatus*. The vegetable oils of mustard, karanj, cedar, apricot and olive are effective against pulse beetle *C. chinensis* in pea seeds. Palm kernel oil is most effective against F1 progenies of *C. maculatus*. Groundnut oil reduced damage and inhibited progeny emergence of *C. maculatus* on cowpea. Application of sunflower, soybean and castor oil to chickpea seeds can control *C. chinensis* (Boeke et al., 2004a)

Seldom, it is important to note that seeds treated with natural plant products and insecticides can show reduced nutritional quality and consumer acceptability. Natural plant products decrease the crude protein content and increase the seed moisture and crude fibre content of treated seeds; but there seems to be no significant adverse effect on organoleptic characters or seed viability. Generally, germination and seed vigour indices decreased with increased concentration of treatment and time of seed storage (Kiran et al., 2012).

27.3 CATALOGUING OF BOTANICALS

Efforts have been made to sharpen the focus on the toxicant and grain protectant activity of essential oils, extracts and their constituents. On the basis of physiological activities on insects, Jacobson (1989) observed the plant families like Annonaceae, Asteraceae, Canellaceae, Labiatae, Meliaceae and Rutaceae as most promising natural grain protectants, in general, and conventionally classified the plant derived compounds into six different groups, viz., **natural grain protectants** insect repellents, feeding deterrents, insect toxicants, growth inhibitors, chemosterilants, and attractants based on the physiological effects in the insects. The use of various plant parts like leaf, bark, seed powder, or oil extracts as admixture to the stored grains have resulted into in reduced rates of seed damage, reduced insect oviposition and suppression of adult emergence in various stored grain insect pests.

27.3.1 Natural Grain Protectants

Starting from very early periods, plant materials have been used as natural protectants of stored grains. Worldwide reports indicate that when mixed with stored grains, leaf, bark, seed powder, or oil extracts of plants reduce oviposition rate and suppress adult emergence of stored product insects, and also reduce seed damage rates (Keita et al., 2001; Bakkali et al., 2008). During 1989, Jacobson noted that the most promising natural grain protectants were generally observed in the plant families, Annonaceae, Asteraceae, Canellaceae, Labiatae, Meliaceae, and Rutaceae.The neem plant is the most well known example and its various parts, specifically, leaves, crushed seeds, powdered fruits, oil, and so forth, have been used to protect stored grain from infestation. The neem oil and kernel powder give effective grain protection against stored grain insect pests like *Sitophilus oryzae, Tribolium cataneum, Rhyzopertha dominica* and *Calloso bruchus chinensis* at the rate of 1 to 2% kernel powder or oil (Pereira and Wohlgemuth, 1982). The neem oil adheres to grain forms uniform coating around the grains against storage pests for a period of 180–330 days (Ahmed, 1994).

Dried leaves of Azadirachta indica have been mixed with stored grains for protection against insects. Azadirachtin is an active principle from the neem plant, which is an effective grain protectant against insect infestation. Rajashekar et al. (2010) reported that root powder extracts of *Decalepish amiltonii* have been mixed with stored grains for protection against various stored grain insect pests. Eighteen species offered protection to wheat up to 9 months without affecting seed germination (Devi and Devi, 2011).

27.3.2 Insect Repellents

The insect repellent chemicals offer protection of stored grains or food commodities from infestation of insect pests with minimal environmental impact through stimulation of olfactory or other receptors. Insect repellent and anti-feedant actions of neem are attributed due to presence of triterpenoid azadirachtin and other related compounds. The neem oil at 0.1 per cent caused effective repellent effect against *T. confusum* on stored corn an also is reported the repellent activity of neem leaf powder in various stored grain pests. Mohiuddin et al. (1987) while testing twelve vegetable oils for toxic and repellent effects against *T. castaneum* reported neem oil at 0.25 per cent as more effective to produce the highest repellency (80.1-100 %) at 60 days after treatment. investigations have shown that the treatments with neem and castor are highly effective as they register significant repellent action on the adult fecundity of pulse beetle. The plant derived extracts, powders, and essential oils were reported as repellent against stored grain insect pests like *C. maculates* and *T. castaneum* (Hassanali et al., 1990).

Wong et al. (2005) compared various plant products, viz., citronella, garlic oil, neemextract, pine oil and pyrethrum for their repellent effects against stored insect pests and found all the products to give positive results. Plant derived repellents are environmentally safe, eco-friendly, effective for pest control, less pesticide residue.

Repellents from plant origins are considered safe in pest control; minimize pesticide residue; ensure safety of the people, food, and environment. The plant extracts, powders, and essential oils from the different bioactive plants were reported as repellent against stored grain insect pests (Koul et al., 2008). For example, the essential oil of *Artemisia annua* is found as repellent against *Tribolium castaneum* and *Callosobruchus maculates* (Tripathi et al., 2000).

27.3.3 Insect Toxicants

Worldwide, research on new toxicants of plant origin has not declined in recent years despite the increased research devoted to the discovery of synthetic insecticides. Information on plant derivatives showed that many plant products are toxic to stored product insects Pascual-Villalobos and Robledo (1998) screened extracts from fifty various wild plant species for their insecticidal activity against *T. castaneum* and revealed four species, viz., *Anabasis hispanica, Senecio lopezii, Bellardia trixago* and *Asphodelus fistulosus* as most promising. Major essential oil constituents from garlic *Allium sativum,* i.e., methyl allyl disulphide and diallyl trisulphide are to be potent toxicant and fumigants against *S. zeamais* and *T. castaneum*. Rahman (1990) revealed nicotine (an active component of *Nicotian atabacum*) as a strong organic poison that acts as a contact-stomach poison having insecticidal properties. This compound is, of course, very toxic to humans as well. The essential oil vapours distilled from eucalyptus, cumin, oregano and rosemary are also reported as fumigants which caused 100 per cent egg mortality of *T. confusum* and *Ephestia kuehniella*. Many species of the genus Ocimum oils, extracts and their bioactive compounds possess insecticidal activities against various insect species (Khanam et al., 1990).

The effects of leaf powders have been tested using *Withania somnifera* L., *Ocimum sanctum* L., *Albizia lebbeck* L., and *Dalbergia sisso* Roxb., as protectants against *C. maculatus* in seeds of stored chickpea. The *W. somnifera* and *O. sanctum* have significant effects on mortality of adults and larvae of *C. maculatus,* and significantly reduced F1 progeny indicating strong deterrence. The possible reasons for the differences may be biochemical response and persistence of the botanicals as protectants of gram against the *C. maculatus* (Sarwar, 2010).

27.3.4 Chemosterilants or Reproduction Inhibitors

Many plant parts, oil, extracts, and powder mixed with grain-reduced insect oviposition, egg hatchability, postembryonic development, and progeny production. Several plant species have been reported as reproduction inhibitors against stored product insects (Copping and Duke, 2007). Different plant derivatives including the essential oils caused mortality of insect eggs of several pests. Numerous ground plant parts, extracts, oils, and vapour also suppress many insects. Various plant parts, powders, oil, extracts when admixed to the grains reduced insect oviposition, egg hatchability, post-embryonic development and progeny production of various storage pests (Dimetry, 2012).

Research has been carried out to identify alternatives to synthetic insecticides to control cowpea weevil *Callosobruchus analis* (F.), population in gram seed (*Cicer arietinum* L.), during storage. The efficacies of three plant materials such as *Nicotiana tabacum, Citrullus colocythis* and *Aloe vera* have been assessed to determine their insecticidal activities against survival of bruchid *C. analis* on seeds of gram varieties, viz., CM-98 and Jubiha-1. These plant materials tested reduced weevil infestation and emergence as compared with untreated control seeds. Seed treatment with *A. vera* followed by *N. tabacum* reduced maximum pest damage over *C. colocythis,* which proved least effective to control *C. analis* population. Consequently, the tested plant materials should be given due consideration for effective gram protection as a component of integrated pest management approach in storage (Sarwar et al., 2012).

27.3.5 Antifeedants or Feeding Deterrents

The chemicals which inhibit feeding or disrupt insect feeding by rendering the treated materials unattractive or unpalatable are called feeding deterrents or antifeedants. Studies on assessment of relative efficacy of plant antifeedants and deterrents against *R. dominica* and *S. cerealella* infestation in stored paddy found neem extract at 1.0 and 0.5 per cent, neem and coconut oils at 1.0 per cent as the highly effective to provide more protection against *R. dominica* for up to 180 days after treatment. A few natural feeding deterrents, which have been characterized, include glycosides of steroidal alkaloids, aromatic steroids, hydroxylated steroid meliantriol, triterpene hemiacetal and others (Mohiuddin et al., 1987). The screening of several medicinal herbs has revealed root bark of *Dictamnus dasycarpus* to possess antifeeding properties against storage insects (Liu et al., 2002).

Khan and Marwat (2002) tested the deterrent effects of leaf, seed and bark powder of neem and oleander (*Nerium oleander*) against *R. dominica* where neem leaves and seeds were found as best to register a highest per cent mortality in the insect.

Laboratory study initiated to analyze various botanical materials such as *Chenopodium album* L. (white goose foot), *Cichorum intybus* L. (chicory) and *Meliolotus parviflora* L. (honey clover) applied at the rate of 2 per cent into wheat grain for the control of *Trogoderma granarium.* The natural protectants used could be useful and desirable tools in stored pest management plan (Sarwar and Sattar, 2012).

27.3.6 Insect Growth Regulators (IGRs)

Insect growth regulators (IGRs) are compounds which interfere with insect metabolism in a manner which affects growth. Plant extracts show deleterious effects on the growth and development of insects and reduce larval, pupal and adult weight significantly, lengthen the larval and pupal periods, and reduce pupal recovery and adult eclosion. The grains coated with plant extracts completely inhibit the development of insect like *Sitophilus oryzae*. Plant derivatives also reduce the survival rates of larvae and pupae and adults' emergence (Singh, 2017).

Development of eggs and immature stages in the grain kernel are also inhibited by plant derivatives. The crude extracts also retard development and cause mortality of larvae, cuticle melanisation, and high mortalityin adults (Obeng-Ofori et al., 1998).

Study reveals that plant extract coated grains are most effective for complete inhibition of development of storage insect such as *S. oryzae*. Plant derivatives are also inhibit the development of eggs and immature stages of insects inside grain kernels. Plant derived products are also found effective to reduce the survival rates of larvae, pupae and the adult emergence (Boeke et al., 2004b).

Plant extracts and essential oils have shown insecticidal activity against household insect pests. Many of these oils have also shown high oviposition and growth inhibitory activity. The neem oil and kernel powder gave effective grain protection against stored grain insect pests like *S. oryzae, T. castaneum, R. dominica* and *C. chinensis* at the rate of 1 to 2% kernel powder or oil (Pereira and Wohlgemuth, 1982). Coconut oil has been found effective against *C. chinensis*, for a storage period of six months, when applied to green gram at 1% (Jayasekara et al., 2005).

The powders of *Rauvolfia serpentina, Acorus calamus* and *Mesua ferrea* are used as a grain protectant against *R. dominica* (Tiwari, 1994). Similarly, volatile compound diallyl disulphide isolated from neem have shown potent toxic, fumigant and feeding deterrent activity against stored grain pests, *S. oryzae* and *T. castaneum* (Koul, 2004). Methyl salycylate from *Securidaca longapedunculata* exhibited repellent and toxic properties against *S. zeamais* and *R. dominica* (Jayasekara et al., 2005). The herbal pellets made from neem seed (*Azadirachta indica*) powder, *sitafal* (*Annona squamosa*) seed powder, asafoetida (*Serula foetida*), chili powder and eucalyptus oil showed significant insecticidal activity against *Bruchus chinensis*. Treatment with *Carissa schimperi* 5% and *Tagetes minuta* 5% was effective against *S. zeamais* and can solve poor farmers' problems by integrating them with other control measures (Chaudhari, 2013).

Bruchids (*Callosobruchus)* are important pests of pulses in countries where legume seeds are often the main source of protein in the human diet and the losses caused by these insects are of major significance. Infestation may start in the pods before harvest and carry over into storage where substantial losses may occur. Delayed harvest of pulses can result in more damage by *Callosobruchus* resulting more seed damage and seed weight loss in fields. Bruchids multiply rapidly in stored condition, giving rise to new generation every month (Sarwar, 2019a). It has been observed that coconut oil provided the best protection on cowpea seeds against *C. maculatus* (Swella and Mushobozy, 2007). Seed damage of 96.15% by *Callosobruchus* has been recorded for cowpea, 88.44% for mung bean (*Vigna radiata)* and 74.22% for moth bean (*Vigna aconitifolia*). Biochemical assays of infested pulses showed that after 20 and 40 days of infestation there is a gradual increase in moisture content, crude fibre and crude fat (Shailja and Meera, 2011). Crude protein initially increased after 20 days, but later showed a slight decrease. Total carbohydrate content decreased. The study also found that seed germination, seedling length and vigour index of cowpea seeds were reduced due to Bruchidae attack.

The studies have been carried out on the repellence activity of plant oils against red flour beetle *Tribolium castaneum* (Herbst) in wheat. The four plant oils, i.e., mustard oil (*Brassica campestris*), coconut oil (*Cocus nucifera*), sesame oil (*Sesamum indicum*) and rocket seed oil (*Eurica sativa*) have been used for the grains treatment. It is observed that plant edible oils affected developmental stages of red flour beetle, that is responsible for causing heavy damage and weight losses to the grains of wheat. However, among oils, the rocket seed oil, statistically gave better results for pest control when compared with other oils and check grain. The overall results presented showed that the minimum adult emergence of red flour beetle was recorded in rocket seed oil followed by mustard oil, coconut oil and sesame oil, whereas, the maximum adult emergence observed in control treatment. Similarly, the minimum pest fecundity is recorded in rocket seed oil followed by mustard oil, coconut oil and sesame oil. On the other hand, the maximum fecundity is observed in control treatment of grain. The minimum grain weight loss is seen in rocket seed oil at its application rate of 0.25 mL per sample. On the other hand, the maximum weight loss is recorded in the control. However, sesame oil is the least effective in protecting grains followed by coconut oil (Hina et al., 2015).

27.4 BOTANICALS AND SEED QUALITY

The grain is stored for sowing purpose in the coming season or year, so the viability and health of seeds is most important. The infestation of seeds due to storage insect pests leads to loss of viability and vigour thereby affecting germination adversely (Rana et al., 2014). Therefore, maintenance of seed quality during storage is essential for successful crop production and integrity maintaining of the seeds that are prone to unpredictable threat of genetic erosion (Barua et al., 2009).

Findings show that botanicals do not have unfavourable effect on germination value of the seeds. The killing or repellent property of various botanicals makes seeds unsuitable for insect pests during storage and enhance the quality parameters (germination, viability, vigor) (Khaire et al., 1992). Botanicals might have the phytotonic effect thereby increase seed quality parameters. Sandeep et al., (2013) recorded higher germination, vigour index and less infestation during storage when maize seeds have been treated with sweet flag rhizome powder.

Various indigenous dried up foliage powders of persian lilac (*Melia azedarach* L.), mudar (*Calotro pisprocera* Willd. R. Br.), sungrass (*Raphanus caudatus* L.), rape (*Eruca sativa* Miller) and castor (*Ricinus comunis L.*) have been tested to contrast their efficacies on the survival of *Caloglyphus* grain mite in wheat grain during storage. The germination test showed that there is no impairment of germination due to treated seeds as compared with untreated grain (Sarwar et al., 2013).

27.5 BOTANICALS IN INTEGRATED PEST MANAGEMENT (IPM)

Integrated pest management (IPM) denotes judicious and effective use of available resources (natural or synthetic) for pest management without harming the ecological balance. Integrated pest management of post-harvest infestation is an effective and most accepted approach (Obeng, 2010). Amruta et al. (2015) recorded effective storage insect control and higher seed quality when treated with botanicals and emamectin benzoate. Insecticidal plant extracts have broad spectrum of action and are easy to prepare at farm level (Azad et al., 2013) from the locally available material. Other researchers, Nukenine et al. (2011) reported that neem based, viz., neem azal and neem seed oil have sufficient efficacy against *S. Zeamais*.

Botanicals are converted to harmless metabolites rapidly and do not help in of build up inherited confrontation in targeted pests and are less detrimental to other beneficial organisms. Thus, different botanicals can be used effectively to treat grains depending upon the locations as activity of botanicals varies across geographical locations; even with differences in storage duration and conditions (Abdalla et al., 2010).

More significant use of natural products in practice is possible if they are compatible with the principles of safe, low toxicity, easy for application, minimal or nil problems with insecticide residues on foods, high efficacy at very low concentrations against stored grain insects, adult pests and their progeny, wide spectrum of efficacy against stored grain insect pests and field pests, low adverse effects on grain handling and quality properties, and acceptable in terms of price (Rozman, 2015).

There is an increasing number of growers who are reaping rewards by placing extra emphasis on the management of their grain storage facilities. It is the combination of practices that provides the real strength to successful storage. Cold winters are an asset in the management of stored product pests, but do not in any way guarantee that the stored product will be pest free. While greater than thirty species of storage pests can attack stored grain, only more than a few species seldom reach to economic injury levels. While several procedures to manage pests are used at storage facilities, before storage those that minimize pest invasion into storage structures include cleaning of bins, and harvesting and loading equipment prior to harvest and after the bin is empty (Sarwar, 2013b; Sarwar, 2004).

Key strategies for integrated storage management include high standard of hygiene for storage and grain handling equipment, minimizing insect-pest breeding sites, monthly checks of grain in storage including those of planting seed, sieving for insects and checking quality, aeration fans fitted to storage and operating by an automatic controller, grain temperature, checking and maintaining 20-23°C in summer and less than 15°C in winter, fumigation when required are carried out in sealable silos, and silos are pressure tested at least once a year. Storage record keeping is a simple system to record details such as grain variety, moisture content, any treatment given, inspection dates, and information on insects found (Siddiqui and Sarwar, 2002; Sarwar, 2019b).

27.6 ULTIMATE ELUCIDATION

Prevention of food losses during post-harvest storage is of paramount economic importance. In the context of grain pests management, botanical pesticides are best suited for use in organic food production in industrialized countries, but can play a much greater role in the production and postharvest protection of foods in developing countries. Botanicals offer many environmental advantages and are emerging as one of the prime means to protect crops and their products. They also possess insecticidal activity to pests, as well as antifungal, antiviral and antibacterial properties against pathogens. An improvement in the understanding of plant allelochemical mechanisms of activity offers new prospects for using these substances in grain protection. This chapter presents different kinds of botanical pesticides from different recourses and their mode of action as well as examining the reasons behind their limited use and the actual grain protection developments involving biopesticides of plant origin for organic or traditional agriculture to keep our environment clean and safe. In general, they act quickly, degrade rapidly and have, with a few exceptions, low mammalian toxicity. In fact, plant like neem (*Azadirachta indica*), has shown excellent results and there are already commercial products in the market made from it.

Pronounced and multiple synergic modes of action substantially reduce or remove the need for synthetic chemically active ingredients, which reduces application and incidental exposure to these chemicals. Efforts should be made scientifically to document the pesticidal plants and to investigate the biocontrol efficacy of the plant products. Farm trials are required to assess the practical applicability of the botanical pesticides. Biosafety studies should be conducted to ascertain their toxicity to humans, animals and crop plants. However, in order to market these products, products containing these active ingredients must be registered for use and must be used in accordance with the provisions of the rules. This chapter helps the producers storing grain on-farm and the commercial elevator operators become familiar with the available methods for managing of storage pests.

REFERENCES

Abdalla, A.S.; Mohamed, E.E.; Abdin, E.M. Insecticidal activities of neem (*Azadirachta indica* A. Juss) seeds under laboratory and field conditions as affected by different storage durations. *Agric. Biol. J. Am.*, 2010, 5: 1001-1008.

Abdullahi, N.; Majeed, Q.; Oyeyi, T.I. Studies on the efficacy of *Vitellaria paradoxa* seed oil on the oviposition, hathability of eggs and emergence of *Callosobruchus maculatus* (F.) (Coleoptera: Bruchidae) on treated cowpea seeds. *J. Entomol.* 2011, 8(4): 391-397.

Ahmed, S. M. Neem (*Azadirachta indica* A. Juss) a safer insecticide potentials and prospects, *Pest Management*, 1994, 47(3): 1-3.

Ali, A., M. Sarwar, S. Khanzada and G. H. Abro Reaction of certain wheat varieties to the action of red flour beetle, *Tribolium castaneum* (Herbst) (Coleoptera) under insectary conditions. *Pakistan Journal of Zoology*, 2009, 41(1): 51-56.

Ali, A., M. Sarwar, S. Khanzada and G. H. Abro. Evaluating resistance of wheat germplasms to attack by red flour beetle, *Tribolium castaneum* (Herbst) (Coleoptera). *Pakistan Journal of Zoology*. 2011, 43(4): 793-797.

Amruta, N.; Sarika, G.; Umesha; Maruthi; J.B.; Basavaraju, G.V. Effect of botanicals and insecticides seed treatment and containers on seed longevity of black gram under natural ageing conditions. *J. App. Nat.Sci.* 2015, 7(1): 328-334.

Azad, A.K.; Sardar, A.; Yesmin, N.; Rahman, M.; Islam, S. Eco-friendly pest control in cucumber (*Cucumis sativa* L.) field with botanical pesticides. Nat. *Reso.*, 2013, 4(5): 6p.

Bakkali, F.; Averbeck, S.; Averbeck, D.; Idaomar, M. Biological effects of essential oils–a review. *Food and Chemical Toxicology*, 2008, 46(2): 446–475.

Barua, H.; Rahman, M.M.; Masud, M.M. Effect of storage container's environment at different storage period on the quality of chilli seeds. *Int. J. Sust. Crop Prod.* 2009, 4(4): 28-32.

Boakye, M.; Mbatchou, V.C.; Khan, M.E. Protecting cowpea seeds (*Vignaunguiculata*) against cowpea weevil (*Callosobruchus maculatus*) with unripe banana peel liquid. *Pharm Pharmaceutical Biotechnology: Current Research*. 2016, 1: 1.

Boeke, S. J.; Barnaud, C.; Van Loon, J. J.A.; Kossou, D. K.; Van Huis, A.; Dicke, M. Efficacy of plant extracts against the cowpea beetle, *Callosobruchus maculatus. International Journal of Pest Management*, 2004a, 50(4): 251-258.

Boeke, S.J.; Baumgart, I.R.; Van Loon, J.J.A, Van Huis. A.; Dicke, M.; Kossou, D.K. Toxicity and repellence of African plants traditionally used for the protection of stored cowpea against Callosobruchus maculatus. *Journal of Stored Products Research,* 2004b; 40: 423-438.

Chaudhari, S.V. Herbal control of stored grain pest *Bruchus chinensis* Linnaeus (Coleoptera: Bruchidae). *Int. J. Innovative Res & Development,* 2013, 2(3): 397-402.

Copping, L. G.; Duke, S. O. Natural products that have been used commercially as crop protection agents. *Pest Management Science.* 2007, 63(6): 524–554.

Deeba, F., M. Sarwar and R. D. Khuhro. Varietal susceptibility of mungbean genotypes to pulse beetle, *Callosobruchus analis* (Fabricius) (Coleoptera: Bruchidae). *Pakistan Journal of Zoology.* 2006, 38(4): 265- 268.

Devi, K. C.; Devi, S. S.Insecticidal and oviposition deterrent properties of some spices against coleopteran beetle, *Sitophilus oryzae, Journal of Food Science and Technology,* 2011, 27: 1–5.

Dimetry, N. Z. Prospects of botanical pesticides for the future in integrated pest management programme (IPM) with special reference to neem uses in Egypt, *Archives of Phytopathology and Plant Protection,* 2012, 45(10): 1138–1161.

Hassanali, W.; Lwande, N.; Ole-Silayo, L.; Moreka, S. N.; Chapya, A. Weevil repellent constituents of *Ocimum suave* leaves and *Eugenia caryophyllata* cloves used as grain protectants in parts of Eastern Africa. *Discovery and Innovation,* 1990, 2(2): 91–95.

Hina, H.K., Sarwar, M.; Lohar, M.K. Repellence activity of plant oils against red flour beetle *Tribolium castaneum* (Herbst) (Coleoptera: Tenebrionidae) in wheat. *International Journal of Animal Biology.* 2015, 1(3): 86-92.

Ileke, K.D.; Odeyemi, O.O.; Ashamo, M.O. Insecticidal activity of *Astoniaboonei* De Wild powder against cowpea bruchid Callosobruchus maculatus F. (Coleoptera: Chrysomelidae) in stored cowpea seeds. *International Journal Biology.* 2012, 4(2): 125–131.

Isman, M. B. Botanical insecticides, deterrents, and repellents in modern agriculture and an increasingly regulated world. *Annual Review of Entomology.* 2006, 51: 45–66.

Isman, M.B. Plant essential oils for pest and disease management. *Crop Protection.* 2000, 19: 603-608.

Jacobson. M. Botanical pesticides: past, present and future. In: Arnason JJ, Philogen B R and Morand P (Eds.) Insecticides of plant origin. *American Chemistry Society Symposium Series*. 1989; 387: 1-10.

Jayasekara, T.K.; Stevenson, PC.; Hall, DR.; Belmain, SR. Effect of volatile constituents from *Securidaca longapedunculata* on insect pests of stored grain. *J. Chemical Ecol.* 2005, 31(2): 303-313.

Keita, S.M. Vincent, C.; Schmit, J.P.; Arnason, J.T.; ´ Elanger, A.B. Efficacy of essential oil of *Ocimum basilicum* L. and *O. gratis simum* L. applied as an insecticidal fumigant and powder to control *Callosobruchus maculatus* (Fab.) (Coleoptera: Bruchidae), *Journal of Stored Products Research*. 2001, 37(4): 339–349.

Kenneth, O.S.; Daniel, A.A.; Ngozi, B. I.; Osa, O.U. Insecticidal and anti-ovipositional activities of the leaf powder of *Jatrophacurcas* (L.) (Euphorbiaceae) against *Callosobruchus maculatus* (F.) (Coleoptera: Chrysomelidae). *Animal Research International*. 2018, 15(1): 2971–2978.

Khaire, V.M.; Kachare, B.V.; Mote, U.N. efficacy of different vegetable oils as grain protectants against pulse beetle, *Callosobruchus chinensis* L. in increasing storability of pigeon pea. *J. Stor. Prod. Res.*, 1992, 28(3): 153-156.

Khan, S.M.; Marwat, A.A. Deterrent/ Repellent effect of different plant parts of neem and kanair against lesser grain borer, *Rhyzopertha dominica* (Fabricius). *Pakistan Entomologist*. 2002, 25: 131-136.

Khanam, L.A.M.; Talukder, D.; Khan, A.R.; Rahman, S.M. Insectidical properties of Royna, *Aphanamixis polystachya* Wall. (Parker) (Meliaceae) against *Tribolium confusum* Duval, "*Journal of Asiatic Society of Bangladesh Science*. 1990, 16: 71–74.

Kiran, B.; Lalitha, V.; Raveesha, K.A. Insecticidal activity of bioactive compound: 2H-Furo [2, 3-H]-1-Benzopyran-2-One, isolated from seeds of *Psoralea corylifolia* L against *sitophilus oryzae*. *Asian Journal of Biomedical and Pharmaceutical Sciences*. 2012, 2(15): 4-6.

Koul, O. Biological activity of volatile dinpropyl disulfide from seeds of neem, *Azadirachta indica* (Meliaceae), to two species of stored grain pests, *Sitophilus oryzae* (L.) and *Tribolium castaneum* (Herbst). *J. Econ. Entomol*. 2004, 97(3): 1142-1147.

Koul, O.; Walia, S.; Dhaliwal, G. S. Essential oils as green pesticides: potential and constraints, *Biopesticide International*. 2008, 4(0): 63–88.

Liu ZL, Xu YJ, Wu J, Goh SH, Ho SH. Feeding deterrents from *Dictamnusdasycarpus* Turcz against two stored-product insects. *Journal of Agricultural and Food Chemistry*. 2002; 50: 1447-1450.

Melo, R.A.; Forti, V.A.; Cicero, S.M.; Novembre, A.D.L.C.; Melo, P.C.T. Use of X-ray to evaluate damage caused by weevils in cowpea seeds. *Hortic. Bras.*, 2010, 28: 472-476.

Mohiuddin, S.; Qureshi, R.A.; Khan, A.; Nasir, M.K.A. Khatri, L.M.; Qureshi, S. Laboratory investigations on the repellency of some plant oils to red flour beetle, *Tribolium castaneum* Herbst. *Pakistan Journal of Scientific and Industrial Research*. 1987, 30: 754-756.

Musa, A. K. Suppression of seed beetle (*Callosobruchus maculatus*) population with root bark powder of *Zanthoxylum zanthoxyloides* (Lam.) Waterm. (Rutaceae) on cowpea (*Vigna unguiculata* (L.) Walp). *Agrosearch*. 2012, 12(2): 196-204.

Ngamo, T.S.L.; Ngatanko, I.; Ngassoum, M.B.; Mapongmestsem, P. M., Hance, T. Persistence of insecticidal activities of crude essential oils of three aromatic plants towards four major stored product insect pests. *African Journal of Agricultural Research*. 2007, 2: 173-177.

Nukenine, E.N.; Tofel, H.K.; Adler, C. Comparative efficacy of Neem Azal and local botanicals derived from *Azadirachtaindica* and *Plectranthus glandulosus* against *Sitophilus zeamais* on maize. *J. Pest Sci.*, 2011, 84: 479-486.

Obeng-Ofori, D. Residual insecticides, inert dusts and botanicals for the protection of durable stored products against pest infestation in developing countries. The 10th International Working Conference on Stored Product Protection: 2010, 474-478.

Obeng-Ofori, D.; Reichmuth, C.H.; Bekele, A.J.; Hassanali, A. Toxicity and protectant potential of camphor, a major component of essential oil of *Ocimum kilimandscharicum*, against four stored product beetles. *International Journal of Pest Management*. 1998, 44: 203-309.

Ojo, D.O.; Ogunleve, R.F. Comparative effectiveness of the powders of some underutilized botanicals for the control of *Callosobruchus maculatus* (Coleoptera: Bruchidae). *Journal of Plant Diseases and Protection* 2013, 120(5/6): 227-232.

Oni, M.O.; Ileke, K.D. Fumigant toxicity of four botanical plant oils on survival, egg laying and progeny development of the dried yam beetle, *Dinoderus porcellus* (Lesne) (Coleoptera: Bostrichidae) *Ibadan Journal of Agricultural Research*. 2008, 4(2): 31-36.

Pascual-Villalobos, M.J.; Robledo A. Screening for anti-insect activity in Mediterranean plants. *Industrial Crops and Products*. 1998, 8: 183-194.

Pereira, J.; Wohlgemuth, R. Neem (*Azadirachta indica* A. Juss) of West African origin as a protectant of stored maize. *Journal of Applied Entomology*. 1982, 25(4): 339-344.

Rahman, M.D.M. Some promising physical, botanical and chemical methods for the protection of grain legumes against bruchids in storage under Bangladesh conditions. In: *Bruchids and Legumes: Economics, Ecology and Coevolution*, 1990.

Rajashekar, Y.; Gunasekaran, N.; Shivanandappa, T. Insecticidal activity of the root extract of Decalepishamiltonii against stored-product insect pests and its application in grain protection," *Journal of Food Science and Technology*. 2010, 47(3): 310–314,

Rana, K.; Sharma, K.C.; Kanwar, H.S. Efficacy of aqueous plant extracts on the seed quality of pea (*Pisum sativum* L.) during storage. *Am.Int. J. Res. For., App. Nat. Sci.* 2014, 6(1): 7-11.

Rozman, V. Control of stored products pests by natural products. Integrated protection of stored products, *IOBC-WPRS Bulletin*. 2015, 111: 295-299.

Sandeep, D.; Chandrashekhar, G.S.; Ranganathswamy, M.; Mallesh, S.B.; Kumar, H.B.H.; Patibanda, A.K. Effect of botanicals on storability of sweet corn (*Zea mays* L. Saccharum) seeds. *Int. J. Plant Prot.* 2013, 6(1): 11-14.

Sarwar, M. Stored grain and stored product mites from Pakistan and Azad Kashmir. *Pakistan & Gulf Economists*. 2004, XXIII (10): 30-31

Sarwar, M. Some possibilities on the effectiveness of plant powders as grain protectants against cowpea weevil, *Callosobruchus maculatus* (Fabricius) Walp (Coleoptera: Bruchidae) infestation in chickpea. *International Journal of Agronomy and Plant Production*. 2010, 1(2): 45-50.

Sarwar, M. Evaluating wheat varieties and genotypes for tolerance to feeding damage caused by *Tribolium castaneum* (Herbst) (Coleoptera: Tenebrionidae). *Pakistan Journal of Seed Technology*. 2009, 2(13 & 14): 94-100.

Sarwar, M.; Ahmad, N.; Bux, M; Tofique, M. Potential of plant materials for the management of cowpea bruchid *Callosobruchus anulis* (Coleoptera: Bruchidae) In Gram *Cicer arietinum* during Storage. *The Nucleus*. 2012, 49(1): 61-64.

Sarwar, M. Studies on resistance discrepancy in seeds of local and exotic lentil (*Lens culinaris* Medikus) genotypes against *Callosobruchus analis* L. (Coleoptera: Bruchidae). *African Entomology*. 2013a, 21(2): 273-279.

Sarwar, M. Development and Boosting of integrated insect pests management in stored grains. *Journal of Agriculture and Allied Sciences*. 2013b, 2(4): 16-20.

Sarwar, M. Extermination of insect pests (Coleoptera: Bruchidae) and damage of stored pulses by different methods in market. *American Journal of Marketing Research*. 2015a, 1(3): 99-105.

Sarwar, M. Distinguishing and controlling insect pests of stored foods for improving quality and safety. *American Journal of Marketing Research*. 2015b, 1(3): 201-207.

Sarwar, M. The killer chemicals for control of agriculture insect pests: The botanical insecticides. *International Journal of Chemical and Biomolecular Science*. 2015c, 1(3): 123-128.

Sarwar, M. Information on activities regarding biochemical pesticides: An ecological friendly plant protection against insects. *International Journal of Engineering and Advanced Research Technology*. 2015d, 1(2): 27-31.

Sarwar, M. A potent folklore of botanical plant materials against insect pests together with their preparations and applications. *Sky Journal of Biochemistry Research*. 2016a, 5(4): 58-62.

Sarwar. M. The rodents (Mammalia: Rodentia) - Gnawing away on stored grains and options for the integrated pest management in stores. *American Journal of Food Science and Health*. 2016b, 2(6): 161-168.

Sarwar, M. Multiple integrated pest management (IPM) and area-wide management (AWM) approaches for insect pests of chickpea *Cicer arietinum* (L.) Walp Crop. In: *Handbook of Chickpeas: Nutritional Value, Health Benefits and Management*, Albert T. Lund and Noah D. Schultz (Eds.). Nova Science Publishers, Inc., New York. 2019 a; pp. 237-263.

Sarwar, M. Multiple insect pests management options for chickpea (*Cicer arietinum* L.) Walp grain with reference to storage. In: *Handbook of Chickpeas: Nutritional Value, Health Benefits and Management*, Albert T. Lund and Noah D. Schultz (Eds.). Nova Science Publishers, Inc., New York. 2019b; pp. 265-302.

Sarwar, M.; Sattar M. Appraisal of different plant products against *Trogoderma granarium* Everts to Protect Stored Wheat- A Laboratory Comparison. *The Nucleus*. 2012, 49(1): 65-69.

Sarwar, M.; Ashfaq, M.; Ahmad. N.; Randhawa, M.A.M. Assessing the potential of assorted plant powders on survival of *Caloglyphus* grain mite (Acari: Acaridae) in wheat grain. *International Journal of Agricultural Science and Bioresource Engineering Research*. 2013, 2 (1): 1-6.

Shailja, R.; Meera, S. Evaluation of qualitative and quantitative losses caused by *Callosobruchus chinensis* to some pulses. *Journal of Entomological Research.* 2011, 35(2): 117-120.A.

Sharifian, I.; Hashemi, S.M.; Darvishzadeh, A. Fumigant toxicity of essential oil of Mugwort (*Artemisia vulgaris* L.) against three major stored product beetles, *Archives of Phytopathology and Plant Protection*. 2013, 46(4): 445-450.

Siddiqui, Q.H.; Sarwar, M. Pre and post-harvest losses in wheat. *Pakistan and Gulf Economist.* 2002, XXI (6): 30-32.

Singh, S. Natural plant products - As protectant during grain storage: A review. *Journal of Entomology and Zoology Studies* 2017; 5(3): 1873-1885

Suleiman, M.; Kankia, I.H.; Sani, Z. Efficacy of *Euphorbia balsamifera* L. and *Jatrophacurcas* L. leaf powders against *Callosobruchus maculatus* F. (Coleoptera: Bruchidae). *European Journal of Zoological Research.* 2014, 3(3): 15-20.

Swella, G.B.; Mushobozy, D.M.K. Evaluation of the efficacy of protectants against cowpea bruchids (*Callosobruchus maculatus*) (F.) on cowpea seeds (*Vigna unguiculata* (L.) Walp.). *Plant Protect. Sci.* 2007, 43: 68-72.

Tesfu, F.; Emana, G. Evaluation of *Parthenium hysterophorus* L. powder against *Callosobruchus chinensis* L. (Coleoptera: Bruchidae) on chickpea under laboratory conditions. *African Journal of Agricultural Research.* 2013, 8: 5405-5410.

Tiwari, S.N. Efficacy of some plant products as grain protectants against *Rhizopertha dominica* (F.) (Coleoptera; Bostrichidae). *Int. J. Pest Mngt.* 1994, 40(1): 94-97.

Tripathi, K.; Prajapati, V.; Aggarwal, K.K.; Khanuja, S.P.S; Kumar, S. Repellency and toxicity of oil from *Artemisia annuato* certain stored-product *beetles. Journal of Economic Entomology.* 2000, 93(1): 43-47,

Ujvary I. Nicotine and other alkaloids. In: *Nicotinoid Insecticides and the Nicotinic Acetylcholine Receptor,* I. Yamomoto and J. E. Casida (Eds.), Springer, Tokyo, Japan. 1999, 29-70.

Wong, K,Y,; Signal, F.A.; Campion, S.H.; Motion, R.L. Citronella as an insect repellent in food packaging. *Journal of Agricultural and Food Chemistry.* 2005, 53: 4633-4636.

28 CHAPTER
Microbial Control of Storage Insect Pests

Muhammad Sarwar

National Institute for Biotechnology & Genetic Engineering (NIBGE), Faisalabad, Pakistan.

Corresponding author: E-mail: drmsarwar64@yahoo.com

ABSTRACT

Many smart methods of insects control depend on microbes, and the more we understand insect-microbe interactions, the more these can be exploited to develop targeted interventions with reduced environmental footprints. This chapter contains information on microbial agents isolated from insect pests and examines their potential for economic pest control. The use of *Bacillus thuringiensis* Berliner against storage pests has considerable potential and its application is a registered method for the control of lepidopteran storage insects. Several groups of protozoa and nematode are of interest as agents for the natural control of insect pests. The use of naturally occurring viruses for the control of insect pests has long been considered to have great potential, as these viruses are pest-specific and relatively safe to vertebrates. Entomopathogenic fungi are the first microorganisms to be used as microbial insecticides and only regarded as suitable for application in humid tropical climates. Therefore, the priority for the future is the adoption of an integrated pest management (IPM) programme, which includes the use of BT where and when appropriate, and it should be closely supervised by the authorities The use of microbial agents for the control of insect pests of stored products is still in its infancy and further research needs to be undertaken. If candidate materials are to be utilized to their best effects, their ultimately approval must be obtained and careful consideration might be given to the role they will play in pest control strategies and public health.

28.1 INTRODUCTION

Losses of grain in storage due to insects are the final components of the struggle to limit insect losses in agricultural production. These losses can exceed those incurred while growing the crop. Stored grain insects are good fliers that move to newly stored grain from fields and from infested grain bins. During the summer and fall, insect infestation is usually on the surface of the grain. In cold weather, insects congregate at the centre and lower portions of the grain and may escape detection until high population numbers are reached. Losses caused by insects include not only the direct consumption of kernels, but also accumulation of frass, exuviae, webbing, and insect cadavers. High levels of this insect detritus may result in grain that is unfit for human consumption (Sarwar, 2015a; 2015b).

Several stored grain or product insect pests (Fig. 28.1) attack and destroy large quantities of stored grains and products during long-term storage in farm bins, grain processing facilities, warehouses, retail stores, and eventually also on the consumer shelves. **Mediterranean flour moth** (*Ephestia kuehniella*) larval stages mainly feed various types of flour. **Indian meal moth** (*Plodia interpunctella*) larval stages infest and feed on different kinds of cereal grains, rice and processed dry foods like pasta, bread and spices. **Sawtoothed grain beetle** (*Oryzaephilus surinamensis*) feeds on broken seeds and seed germs. **Mealworm** (*Tenebrio molitor*) larvae feed on flour and cereals.**Red flour beetle** (*Tribolium castaneum*) feeds on flour, cereal grains and dried food products like pasta, biscuits etc. **Warehouse beetle** (*Trogoderma variabile*) larvae feed on dried cereal grains and food products such as noodles and spaghetti, and dried spices (Siddiqui and Sarwar, 2002).

Fig. 28.1 Stored grain or product insect pests.

Armies of microbes and their powers enhanced in some cases by genetic engineering, are unleashed in fields and laboratories in an effort to control destructive insect/pests without resorting to chemical pesticides.Entomopathogens are microorganisms that are pathogenic to arthropods such as insects, mites and ticks. Several species of naturally occurring bacteria, fungi, nematodes and viruses infect a variety of arthropod pests and play an important role in their management. Some entomopathogens are mass produced in vitro (bacteria, fungi and nematodes) or in vivo (nematodes and viruses) and sold commercially. In some cases, they are also produced on small scale for non-commercial local use. Using entomopathogens as biopesticides in pest management is called microbial control (Sarwar, 2015c), which can be a critical part of integrated pest management (IPM) against several pests (Sarwar, 2004; 2013).

Some entomopathogens have been or are used in a classical microbial control approach where exotic microorganisms are imported and released for managing invasive pests for long-term control. The release of exotic microorganisms is highly regulated and is done by government agencies only after extensive and rigorous tests. In contrast, commercially available entomopathogens are released through inundative application methods as biopesticides and are commonly used by farmers, government agencies and home owners. Understanding the mode of action, ecological adaptations, host range and dynamics of pathogen-arthropod-plant interactions is essential for successfully utilizing entomopathogen based biopesticides for pest management in agriculture, horticulture, orchard, landscape, turf grass, urban environmentsand storage insect pests (Ramazan, 2016).

Several entomopathogens, when inundatively introduced into a variety of habitats, can provide effective long-term to short-term control. The most propitious integration of pathogens, predators, insect growth regulators and conventional insecticides may provide long-term control of serious storage insect pests.

Identification of alternatives to synthetic contact insecticides for the protection of durable foodstuffs has become a focus for research worldwide during the last several years. The aim has been to find products to replace or enhance current control methods which rely on fumigation and insecticide application.

28.2 MICROBIAL CONTROL AGENTS FOR SUPPRESSION OF STORED PRODUCT INSECTS

Different types of organisms including pathogenic microorganisms such as viruses, bacteria, fungi, protozoa, and nematodes have been used as biocontrol agents. Most of the field tests have been conducted with viruses and bacteria. Rickettsiae, fungi, protozoa and nematodes have not been extensively utilized, mainly because of difficulties inherent in their use or because of their extreme dependence on environmental conditions. Experimentations have proved that microbial control, or the use of pathogenic microorganisms, is an integral part of applied insect control. Microbial control is of special value when other methods have been found inadequate, for example, when insecticidal residues are toxic to man, insect parasites and predators. Some of the potential candidates for biocontrol of storage insect pests are discussed here.

28.2.1 Insect Bacteria

Among the bacteria, the entomogenous sporeformers have received increased interest in recent years because of their potential for insect control. These spore formers may be roughly divided into two general groups: those related to the milky disease organism, *Bacillus popilliae* Dutky, and those related to *Bacillus cereus* Frankland & Frankland (noncrystal bearers) and *Bacillus thuringiensis* Berliner (crystal bearers).One of the few nonspore formers used in field application is *Pseudomonas aeruginosa* (Schroeter) Migula (Sarwar, 2016a).

28.2.1.1 *Bacillus Thuringiensis Berliner*

A Gram-positive flagellated rod-shaped bacterium *B. thuringiensis* (BT) produces a parasporal crystal during sporulation. When ingested, this proteinaceous crystal is responsible for the toxic effect in susceptible Lepidoptera, Diptera and Coleoptera larvae.The histological symptoms in the infected host are enlargement, distension and disintegration of the midgut epithelial cells. Pathogenic spores germinate in the gut and the multiplying vegetative bacterial rods invade the haemocoel producing toxaemia and septicaemia. The external signs of the disease are larval sluggishness, flaccidity and dark brown spots on the cuticle. Cadavers which have turned dark brown are filled with bacterial spores. Commercial preparations of BT can be applied as a wettable powder, liquid formulation or dust for controlling moth infestations, particularly *Ephestiacautella* and *Plodia interpunctella* in stored grains and soya beans (Dales, 1994).At least eight varieties of BT have been recovered from stored-product moth larvae following natural infections (Table 28.1).

Table 28.1 Varieties of BT recovered from storedproducts moth larvae

Insect Larva source	BT variety
Corcyra cephalonica	var. galleriae
Ephestia cautella	var. kenyae
E. elutella	var.kurstakimorrisoni
E. kuehniella	var. kurstakimorrisoni
E. kuehniella	var. thuringiensis
Plodia interpunctella	var. galleriae
P. interpunctella	var. subtoxicus
Nemapogon granella	var. tolworthi

The study aimed at testing the effect of bacterial poison *B. thuringiensis* in adults and larvae of *Trogoderma granarium* (Fig. 28.2) and in two directions in laboratory and in storage, showed a clear effect of bacterial toxin and its different concentrations in larval instars of the insect , The highest mortality rates in larval instars (first, second, third, fourth and fifth) is at 6 g / kg wheat which reached (66.7, 63.3, 53.3,43.3 and 36.6%), respectively, and the highest percentage of mortality is in the larvae of the first instar and the lowest in the larvae of the fifth instar. The highest mortality rate in insect adults is 6 g/kg of bacterial poison, which is 23.4%. The results showed a significant effect of bacterial toxin in the first generation that resulted from the treatment of first instar larvae found also the lowest percentage of adults and the lowest loss of grain weight and the lowest percentage of grain damage are at 6 g/kg wheat weight (70.0% and 26.7% and 0.03 g / 5 g wheat and 10.3%), respectively, while the results of stored study after three months of treatment indicated the average number of adult emergence of the insect, and loss of the weight of the grain and the lowest proportion of damage in cereals is at concentration of 6 g/kg wheat, which is (104 adult and 2.09 g / 250 g and 0.7%), respectively. The study results showed that bacterial toxicity of *B. thuringiensis* can be used to protect the stored grain from the infection of the storage insects, especially the grains for planting (Alaa et al., 2018).

Fig. 28.2 *Trogoderma granarium.*

A study has reported the biological control of stored product insect pest, *Callosobruchus maculatus* (Fig. 28.3) using the entomopathogenic bacteria *B. thuringiensis*. A significant delay in the larval, pupal and total development period of *C. maculatus* has been observed after treatment with *B. thuringiensis*at 4 × 10^8 cells/ mL. Furthermore, *B. thuringiensis* was highly effective in the control of *C. maculatus* and produced 100% mortality at 4 × 10^8 cells/ mL. The LC_{50} value is estimated to be 3 × 10^7 cells/mL. In addition, a significant decrease in the activity of mid-gutα-amylase, cysteine protease, α & β-glucosidases, lipase, glutathione S-transferase and lactate dehydrogenase is observed after treatment with *B. thuringiensis* at 4 × 10^8 cells/ mL (Malaikozhundan and Vinodhini 2018). This study concludes that *B. thuringiensis* is more effective against *C. maculatus* and could be used as a potential biological control agent in the management of stored product insect pests.

Fig. 28.3 *Callosobruchus maculatus.*

The BT is rated as a safe microbial insecticide which is harmless to vertebrates including man and also harmless to beneficial insects such as bees. However, endocardial and myocardial haemorrhages, and lesions in the heart, liver and lung abnormalities have been observed in sheep fed on maize treated with 250 and 500 mg of formulation/kg. Histological examination revealed the presence of bacterial BT rods in the infected organs. It is thought that an inert material in the formulation may have created a route of entry for the bacteria and further studies using pure spores are recommended. High residues of BT on grain are not thought to pose toxic or physical problems as grain processing eliminates most of the spores (Subramanyam and Cutcomp, 1985). A labourer who, when applying BT for the control of Lepidoptera, accidently splashed the formulation into his eye resulting a corneal ulcer developed which required treatment with gentamicin to cure the infection. So, eye protection is recommended as a safety procedure for operators (Samples and Buettner, 1983).

Resistance to BT has been developed in the laboratory amongst strains of *Plodia interpunctella* (Fig. 28.4) and *Ephestia cautella* (Fig. 28.5) reared on a treated diet. Chiang et al. (1986) examined the defense reaction of midgut cells of *Corcyra cephalonica* (Fig. 28.6) during an infection using scanning and sectioning techniques. They found that following an infection, the epithelial cells become loose, the columnar cells swell and new cells develop in the basal portion of the epithelium. A protective mucous layer covers the surface of the epithelium cells and thus protects the new cells from toxic attack. These defense mechanisms of the midgut cells prolonged the life span of the infected larvae. Thus, the use of BT where and when appropriate, must be closely supervised by the authorities for resistance monitoring.

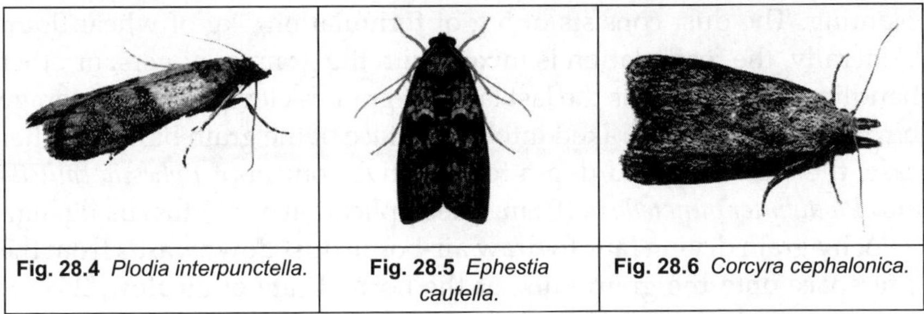

Fig. 28.4 *Plodia interpunctella.*	**Fig. 28.5** *Ephestia cautella.*	**Fig. 28.6** *Corcyra cephalonica.*

28.2.1.2 Pseudomonas Syringae

The *Pseudomonas syringae* (strain 31 a) is a common foliar bacterium isolated from maize leaves. It can be grown under conditions which maximize its ice nucleating activity and then concentrated, freeze dried and killed with electron beam irradiation. These preliminary trials indicate that *P. syringae* has potential for reducing the cold-hardiness of insect pests of stored products specially *Cryptolestes ferrugineus* (Fig. 28.7). Increasing the concentration of *P. syringae* raised the super cooling points of the treated insects; this reduced their tolerance of sub-zero temperatures and, therefore, increased cold-induced mortality (Fields, 1991).

Fig. 28.7 *Cryptolestes ferrugineus.*

28.2.1.3 Other Bacterial Species

Other bacterial species isolated from the gut of *Tribolium castaneum* have been identified and examined for their pathogenicity. The four asporogenous species, *Enterobacteraero genes, E. cloacae, Proteus vulgaris* and *P. mirabilis,* and two sporeformers, *Bacillus subtilis* and *B. cereus,* have been administered orally to *T. castaneum* larvae by the diet dilution technique (at 0.01 mL of 1.0 optical density units). The three bacteria *E. aerogenes, E. cloacae* and *B. cereus* are considered to be potential control agents (Kumari and Neelgund, 1985).

Fermentation processes have been widely used for the commercial production of bacteria. Initially, to ensure product consistency the growth medium and fermentation conditions must be carefully defined (Daoust, 1990). Dust and wettable powder formulations can be used for either crop seed or stored food grain. The liquid formulations are easily applied in water, but their use is frequently limited to seed for planting. The dust consists of 5 g of formulation/ kg of wheat flour. Generally, the formulation is mixed with the grain in augers, or other handling equipment, as the last layer of grain is elevated in the storage bin; alternatively, it is raked into the surface of the grain bulk. In either case, the recommended depth for effective control of *Ephestia cautella* and *Plodiainter punctella* is 10 cm. The application of BT dust using high velocity grain drying fans to draw air-borne dust downwards from the overspace onto the grain bulk, at the normal rate of air flow, 25% of the dust penetrates 2.5-12.5 cm into the corn and prevents infestation. The method may be more effective in commodities with large kernels or pods (McGaughey, 1987).

28.2.2 Insect Viruses

Of the five general groups of virus diseases, nuclear and cytoplasmic polyhedroses, granuloses, polymorphic inclusion viroses, and the non-inclusion viruses, only the viruses of the granuloses and nuclear polyhedroses have been utilized extensively in microbial control. The group of viruses which has attracted the most attention is in the family Baculoviridae. The Indian meal moth *Plodiainter punctella*, is attacked by a granulosis virus. Interest has shown in developing of the granulosis virus from *P. interpunctella* (IMMGV) for use as a protectant in stored dried fruits and nuts. However, semipurified homogenates of larvae infected with the virus lose their infectivity through time, particularly at temperatures above 32°C. A freeze-dried, powdered formulation of IMMGV with a relatively long shelf life is then developed which could be applied as a spray or dust (Cowan et al., 1986).

Vail and Tebbets (1991) developed a modified rearing diet for *P. interpunctella* omitting glycerol and honey; 10-day old larvae inoculated with the granulosis virus and incubated at 26.7°C for 10 days are able to survive for at least one generation on this diet. The diseased larvae and diet are then harvested, homogenized in sterile distilled water, freeze-dried, and milled to a particle size which could be applied through a spray nozzle or as a dust. The LC_{99} for this formulation is 14 µg of granulosis virus formulation/g of diet. It is estimated that at an application rate of 14 µg/g, a 200 g production lot would be sufficient to treat 14,000 kg of commodity. Commercial trials are carried out on a dried fruit packing line in which the same formulation is applied to raisins. It is found that at 27°C, insect control could be achieved for up to 6 months; however at higher temperatures (32-38°C), efficacy is reduced after 1-3 months of storage.

Viruses can be applied at very low doses as they persist and multiply in the pest population to produce long-term control. However, viruses mostly affect the larval stage and are relatively slow acting. A major criticism of virus research to date encountered is that there are many obstacles to the use of viruses on stored food. One of these has been the ethical question of treating material destined for human or animal consumption with insect pathogenic viruses. The general public often fears microbial agents because of their association with ailments such as the common cold, stomach flu, AIDS, etc. (Falcon, 1990). A prerequisite for the use of microbial control would be the implementation of a public awareness campaign to inform the community for the benefits of this type of control system.The success already achieved with the *P. interpunctella* granulosis virus indicates that if other pathogenic viruses can be isolated from the major insect pests of durable foodstuffs, the method could have considerable potential(Sarwar, 2016b).

28.2.3 Insect Nematodes

Entomopathogenic nematodesalso called insect-parasitic nematodes *Steinernema* and *Heterorhabditis* species are commercially available for the control of stored product insects, including those belonging to orders Coleoptera and Lepidoptera. For example, **entomopathogenic nematodes** *Steinernema carpocapsae* use ambush foraging called "sit and wait" strategy to attack highly mobile insects including stored-product pests. After application, infective juveniles of *S. carpocapsae* generally remain near or at the surface of the stored products.When infective juveniles of *S. carpocapsae* sense that there is an insect host passing by them, they attack and infect it by standing on their tails (behaviour called 'nictation') and jumping on the host (Tradan et al., 2006; Fayyaz and Javed, 2009).

Ambush foraging **entomopathogenic nematode** *Steinernema carpocapsae* have a capacity to cause over 85% larval mortality of Indian meal moths, mediterranean flour moths, mealworms and red flour beetles (Ramos-Rodriguez et al., 2006).

Entomopathogenic nematodes such as *Heterorhabditis bacteriophora, Heterorhhabdtits megidis, Steinernema glaseri* and *Steinernema kraussei* are considered cruiser nematodes because they generally move actively in search of hosts and can easily find and attack their insect hosts that are hiding deep in the soil, or in case of stored products hiding in cracks and crevices and under perforated floors. Cruiser nematodes never nictate but use carbon dioxide released by insect hosts as cues to attack them. Cruiser entomopathogenic nematodes, *H. bacteriophora and H. megidiscan* kill larvae of Indian meal moth (Mbata and Shapiro-Ilan, 2005; Sarwar et al., 2020).

Some **entomopathogenic nematodes** such as *Steinernema feltiae* and *Steinernema riobrave* have adapted a strategy in between ambush and cruise strategies called intermediate strategy to attack both the mobile and sedentary/less mobile insects at the surface or deep in the soil and in case of stored products, pests that hiding in cracks and crevices and under perforated floors or remaining at the surface of the product (Athanassiou et al., 2010).

Intermediate foraging **entomopathogenic nematode**, *S. riobrave* have a potential to kill over 65% larvae of Indian meal moths, mediterranean flour moths, saw toothed grain beetles (Fig. 28.8), mealworms, red flour beetles and warehouse beetles. Other intermediate foraging **entomopathogenic nematode** *S. feltiae* can cause over 90% larval mortality of only indian meal moths, mediterranean flour moths, red flour beetles (Ramos-Rodriguez et al., 2006) and over 79% larval mortality of the confused flour beetle *Tribolium confusum* (Athanassiou et al., 2008).

Fig. 28.8 Sawtoothed grain beetles.

28.2.4 Insect Protozoa

Protozoa offer considerable promise in microbial control because of the large number of parasitic species found on insects. Several groups of protozoa are of interest as agents for the natural control of insect pests. Quite a few pathogenic protozoan species have been isolated and identified from insect pests of durable foodstuffs. Coccidia *Adelina tribolii* infects several stored product pests and causes epizootics in laboratory and natural populations of *Tribolium confusum* (Brooks, 1988).Eugregarines are frequently encountered as commensals in the digestive tract of insects. Of those thought to be potentially pathogenic to their hosts, Ascogregarina species have received the most attention. An *Ascogregarina bostrichidorum* has been isolated from *Prostephanus truncatus* and examination of the gut of heavily infested larvae revealed that they contained masses of cysts and spores (Purrini and Keil, 1989).

Neogregarines occur naturally in Lepidoptera, Coleoptera and Orthoptera, and have been considered potential control agents against species belonging to these three orders.A *Farinocystis tribolii* is parasite of *Tribolium destructor* (Fig. 28.9), *T. molitor* (Fig. 28.10), *T. castaneum* (Fig. 28.11) and *T. confusum* (Fig. 28.12). Farinocystis species have also been isolated from *P. truncatus* (Schulz and Laborius, 1987). An infection can be spread by the dispersal of spores from dead larvae during the handling or processing of an infested commodity, or by adults feeding on the bodies of the dead larvae. Infection by *F. tribolii* results in a slow decline of Tribolium species. A *Mattesia trogodermae* has been isolated from *Trogoder magranarium*. This protozoan is cosmopolitan, males surface contaminated with spores will inoculate females while mating. The efficiency of spore transfer can be increased by releasing a natural female sex pheromone at the site where the males become contaminated. It is non-pathogenic to vertebrates, and it is relatively easy to produce and isolate in usable quantities (Henry, 1981). Mattesia species have also been detected in *P. truncatus* collected from farm stored maize.

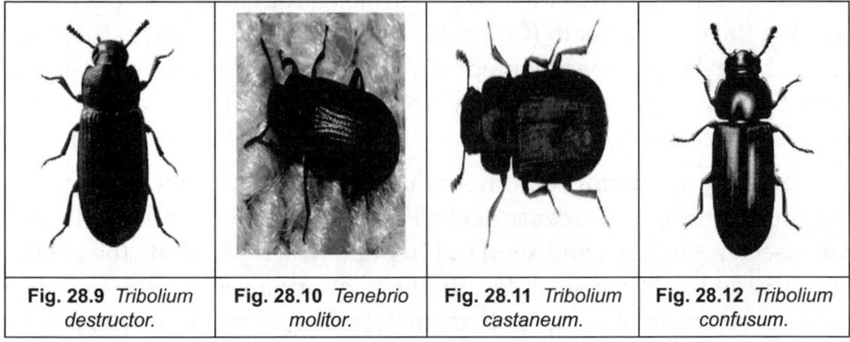

Fig. 28.9 *Tribolium destructor.* **Fig. 28.10** *Tenebrio molitor.* **Fig. 28.11** *Tribolium castaneum.* **Fig. 28.12** *Tribolium confusum.*

Microsporidia Nosema species have been isolated in *P. truncatus* (Fig. 28.13) (Schulz and Laborius, 1987). Most of published articles however, relate to the use of *Nosema whitei* in Tribolium species. It also reduced fecundity and fertility and increased adult mortality in infected insects (Khan and Selman, 1988).

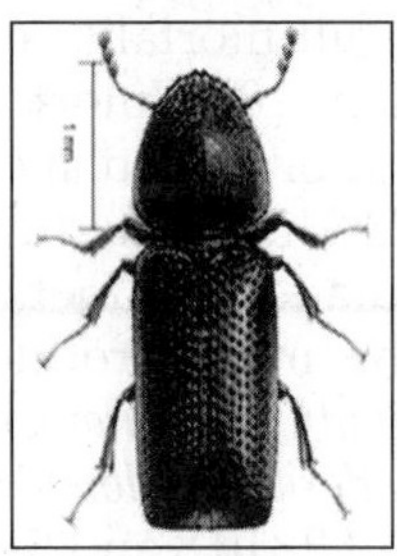

Fig. 28.13 *Prostephanus truncatus.*

Microbial control using protozoa has advanced by adopting many of the principles and methods used in chemical control and in the parasite-predator branch of biological control. The use of microbial agents for the control of insect pests of stored products is still in its infancy and further research needs to be undertaken.In spite of the costs, and the risks and time involved in researching in microbial control, a few successes have been in producing useful products for protectants of durable foodstuffs (Sarwar, 2016c).

Control recommendations of stored product insects rely on broad spectrum chemical insecticides. Due to regulatory and environmental concerns, effective alternatives for control must be sought for grains protection in conventional and organic systems. The results provide evidence that microbial biopesticides can substantially reduce stored product insect infestations in organic and non-organic systems.

28.2.5 Insect Fungal Pathogens

The entomopathogenic fungi are promising for the control of stored product insects as environmentally safe and with low mammalian toxicity. The fungus attaches and penetrates through the insect's cuticle, causing the insect's death (Cox and Wilking, 1996). Entomopathogenic fungi have been experimented with success against several stored product insect species in both laboratory and field tests (Sabbour and Abd El-Aziz, 2010; Sabbour et al., 2012; Kavallieratos et al., 2014).

Naturally occurring entomopathogenic fungi do not have to be ingested by the insect to cause death. Fungal spores stick to the surface of the insect, germinate, and send out hyphae which penetrate the cuticle and invade the haemocoel. Death either occurs rapidly, due possibly to the production of complex toxic metabolites, or more slowly, due to hyphal proliferation and disruption of organs. The invading fungus then sporulates and re-enters the ambient environment to establish subsequent infections (Sarwar, 2016d).

The application of *Beauveria bassiana* to glasshouse and field crops has been studied extensively. Searle and Doberski (1984), investigated the use of *B. bassiana* isolated from a soil sample against *Oryzacphilus surinamensis*. The highest adult mortality occurred at 25°C and 100% r.h.,and under these conditions 100% mortality has occurred within 13 days.It is concluded that in grain stored at or below the recommended moisture content of 14% (70% r.h.), the fungus would be unlikely to control *O. surinamensis* populations. A method of surface fermentation has been developed for the mass production of *B. bassiana* using *Sitophilus zeamais, Oryzeaphilus surinamensis, Tribolium castaneum, Sitophilus granarius*and *Galleria mellonella* (Fig. 28.14). The LC_{50} appeared to be in the range of 2×10 8 and 5×108 conidia/ml (Frydocva et al., 1989).

Fig. 28.14 *Galleria mellonella.*

Investigations carried out to determine inoculation spray rates for *Ephestia cautella* larvae in stored dates indicate that 300 000-400 000 spores/cm^3 are needed to produce 96-98% mortality (Jassim et al., 1988). Studies by Schulz and Laborius (1987) on natural fungal parasites of *Prostephanus truncatus* strains showed that all the fungal isolates belonged to the Deuteromycotina; most are species of Aspergillus and Penicillium and included the mycotoxigenic *A. flavus*.

Entomopathogenic fungus *Metarhizium anisopliaeis* is with a worldwide distribution. It can be cultivated on both solid and liquid sterile media. The product is very effective against Coleoptera and Lepidoptera, and has been tested against pests of ornamental crops (Reinecke et al., 1990). Rodrigues and Pratissoli (1990), carried out small-scale laboratory trials to evaluate the pathogenicity of *Beauveria brongniartii* and *M. anisopliae* for *Sitophilus zeamais* and *Acanthoscelides obtectus*. Adult insects are dipped in conidial suspensions (108 conidia/mL), returned to maize or bean host, respectively, and mortality caused in either species.

The avermectins are a mixture of natural products produced by a soil actinomycete *Streptomyces avermitilis*. The avermectins have nematicidal, acaricidal and insecticidal activates. Their insecticidal properties were first demonstrated in laboratory assays against *Tribolium confusum*and selected for crop protection purposes (Lasota and Dybas, 1991). Beeman and Speirs (1984) carried out tests with avermectin B1 against a range of storage pests. It caused 100% mortality in parent adult *Sitophilus oryzae* (Fig. 28.15), *Rhyzopertha dominica* (Fig. 28.16) and *Oryzacphilus surinamensis* exposed to a dose of 320 ppb on wheat.

Fig. 28.15 *Sitophilus zeamais.*

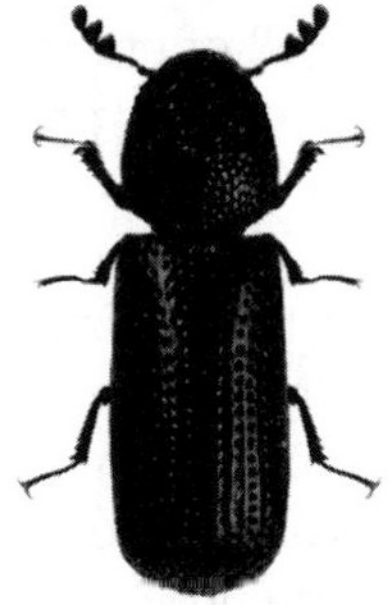

Fig. 28.16 *Rhyzopertha dominica.*

Both silica gel and diatomaceous earths kill insects by absorbing the oily or waxy outer cuticle layer by direct contact. When the thin waterproof layer of the epicuticle is lost then the insect loses water and dies. Several diatomaceous earth formulations and silica have been successfully evaluated as grain protectants against different stored grain pests (Iatrou et al., 2010; Athanassiou et al., 2013).

The occurrence of entomopathogenic fungi isolated from stored grain insect pests sampled from various geographical regions investigated a total of 25,720 insects from six different species evaluated, and 195 isolates from 24 different fungal species recovered. These included the Ascomycetes *Beauveria bassiana sensulato* (Balsamo) Vuillemin (Hypocreales: Clavicipitaceae), *Metarhizium anisopliaesensulato* (Metschnikoff) Sorokin (Hypocreales: Clavicipitaceae), *Purpureocillium lilacinum* (Thorn) Samson (Hypocreales: Ophiocordycipitaceae), and *Lecanicillium attenuatum* (Zare and W. Gams) (Hypocreales: Clavicipitaceae). The cadavers of red flour beetle *Tribolium castaneum* (Herbst.) (Coleoptera: Tenebrionidae) are significantly infected with the fungi followed by rice weevil *Sitophilus oryzae* (L.) (Coleoptera: Curculionidae), lesser grain borer *Rhyzopertha dominica* (F.) (Coleoptera: Bostrichidae), rusty grain beetle *Cryptolestes ferrugineus* (Stephens) (Coleoptera: Cucujidae), and cowpea weevil *Callosobruchus maculatus* (F.) (Coleoptera: Bruchidae); however, the least are recovered from khapra beetle *Trogoderma granarium* (Everts) (Coleoptera: Dermestidae). The geographical attributes (altitude, longitude and latitude) greatly influenced the occurrence of entomopathogenic fungi with highest number of isolates found from > 400 (m) altitude, 33-34′ N latitude and 73-74′ E longitude (Wakil et al., 2013). The findings of these current surveys clearly indicated that the entomopathogenic fungi are widely distributed in the insect cadavers, which may later be used in successful Integrated Pest Management programmes.

The red flour beetle *Tribolium castaneum* is found wherever grains or other dried foods are stored. In the research, there is isolated the pathogenic KB512 of entomopathogenic fungi *Beauveria bassiana* from soil and insects, which produced aerial and submerged conidia and blastospores in laboratory conditions. The best conditions for the production and utilization of spore suspension to spray the larvae of *T. castaneum* have been investigated. One hundred and eighty isolates that naturally infected by *T. castaneum* are reared and pathogenicity test is carried out with direct spray. To bioassay the isolates, the concentrations of the spore suspension have been prepared as follows: 1×10^6 and 1×10^8 conidia/ mL, and the pest sprayed by aerial conidial suspensions. After spraying on the pests, the plates incubated at 25 ± 1 °C and 80% relative humidity. Then, the treated pest has been found every day for the fungal growth and mortality (Zamani et al., 2013).

The efficacy of three fungal pathogens *Isariafarinose* (Holmskjold) Fries (= *Paecilomyces farinosus* (Holmskjold) Brown and Smith),

Nomuraea rileyi (Farlow) Sampson (Ascomycota: Hypocreales) and *Lecanicillium lecaniiz* are and Gams (Ascomycota: Eurotiales) (= *Verticillium lecanii* (Zimmermann) alone and their combination with natural diatomaceous earth and silicagel evaluated against *Tribolium castaneum* and *Tribolium confusum*. Results showed that modified diatoms with calcium hydroxide and modified diatoms with sodium hydroxide are the most effective treatments against the two tested insects and achieved the highest mortality percentages. The fungus *I. fumosoroseais* the most effective alone against *T. castaneum*. Diatomaceous earth and silica gel treatments strongly enhanced the potency of the tested fungi *I. fumosorosea* and *N. rileyi*. In most cases, diatomaceous earth combinations with tested fungi has synergistic effects, while in *T. castaneum* modified diatoms with aluminium hydroxide decreased the efficacy of *V. lecanii*. Both silica gel and diatoms protected wheat better. The egg production is highly suppressed by combination of calcium - diatoms followed by sodium - diatoms with tested fungi in comparison to untreated control. The combination of calcium - diatoms /*I. fumosorosea* strongly suppressed the number of deposited eggs of *T. castaneum* (87.5±9.6 eggs/female) in comparison to untreated control (277.0 ± 5.9 eggs/female). The most effective diatomaceous earths modifications are calcium - diatoms and sodium - diatoms with insecticidal, repellency or ovicidal effects against *T. castaneum* and have synergistic effects on the potency of tested fungi (Sabbour, 2014).

Among the fungi, the genus Beauveria appears to offer the most promise in microbial control because of its rather high pathogenicity for a large number of hosts. But, fungi have seen rather limited field application in recent years, apparently because of their great dependence on proper humidity conditions.

The major drawbacks to the use of fungi for insect control are thought to be their poor stability in storage situations and high dependence, for efficacy, on climatic conditions in agricultural situations (Kirschbaum, 1985). A successful rate of their infection depends on spore germination. This requires optimum temperatures and relative humidities in excess of 80%. Consequently, fungi are only regarded as suitable for application in humid tropical climates and greenhouse situations. If the problems of temperature and moisture necessary for conidial discharge and spore germination could be overcome, use of fungi might be a practical method for controlling of insects in stored products. Overall, fungal control of storage pests may warrant further investigation.

28.3 METHODS OF APPLICATION

Insect pathogens can be utilized in a variety of pest management approaches, from inundative release to augmentation and classical biological control, and microevolution and the consideration of evolutionary principles can potentially influence the success of all these strategies. Comparative analyses of pathogen population structure, virulence, and host resistance over time are required to elucidate the evolutionary dynamics of microbial control systems. Entomopathogens are distinguished from other means of insect biological pest control primarily by their methods of application. For the most part, insect pathogens are applied using an inundative release strategy. That is, they are applied in large numbers onto intermediate to high densities of pest populations, with the expectation of 'immediate' pest control. Immediate means that pest suppression does not rely on the long-term reproduction or establishment of the pathogen, although in reality it can take several days for the pathogens to replicate and kill their host. Insect pathogens, although primarily applied as short-term pest control agents, could benefit from consideration of evolutionary principles in several areas. Given the large natural variation of insect pathogens, selection for particular traits is an approach that could yield benefits if it is clear what traits are required for a particular strategy (Cory and Franklin, 2012).

Ideally, bait formulated with an attractant would be attractive both physically and chemically stored product insects, and it should contain a feeding stimulant to ensure ingestion and contain enough pathogen to be lethal. Another method of effectively distributing a pathogen among stored product insects would be to provide a pheromone baited or light baited device with an open reservoir containing a pathogen such as *B. thuringiensis*. The insects attracted by the pheromone or light would become dusted with the pathogen and would distribute it within the insect population and habitat.

There are four major ways to apply pathogens for insect control: a) applied alone as a spray, dust, or bait; b) introduction and colonization into an insect population; c) application in combination with insecticides, both compatibly and synergistically; and d) application combined or coordinated with insect parasites, predators, and other microorganisms. Thus far, the first two methods have been used most frequently, and only a few attempts have been made in combination with insecticides, parasites, or other microorganisms.

Most of the present-day conventional sprayers and dusters, and mist blowers have been found suitable for the application of insect pathogens, especially viruses and bacterial spores. Insect pathogens are usually applied in their resistant state or stage: the viruses within their inclusion bodies (polyhedra or capsules); bacteria, fungi and protozoa as spores; and the nematodes in their resistant dauer stage. Nematodes and protozoa are apparently capable of withstanding the usual pressures of sprayers. Candidate materials should have included plants and plant extracts, inert dusts, insect growth regulators and microbial insecticides.

Entomopathogens, viz., fungi, bacteria, virus, nematodes and protozoa have an important place in the biological control because they have a wide host range, are harmless to the environment and humans, and could be applied with conventional sprayers. They can be used more against stored product pests with the development of new biotechnical methods such as collecting pests in some stations to meet them with entomopathogens (Canhilal, 2016). Naturally occurring biological control agents must be patented very carefully, however, there is a suggestion, that federal governments could establish central microbial pesticide development centres which could mass produce and formulate viruses into standardized experimental products available for distribution to cooperating insect specialists (Huber, 1990).

The grain protectants currently approved are likely to be the main agents of insect pest control in the foreseeable future. The number of compounds in use may even be reduced as regulations change. However, in order to extend the options, alternative materials with desirable properties need to be evaluated for their potential as grain protectants before the need for new materials becomes critical. Also, since most stored product infestations comprise several insect species, there would be a need for several different control techniques(Sarwar, 2019).

28.4 CONCLUSION

Biological control of stored grain pests may prove effective, if used in appropriate manner, time and space. The concept of microbial control of insects by use of viruses, bacteria, fungi, nematodes and protozoa have been developed for a broad range of applications, but their use as grain protectants requires further investigation.

Recently, their introduction has attracted greater commercial interest due to developments in technology and closer collaboration between the disciplines of molecular biology, genetics, microbiology, and protein and nucleic acid chemistry.Novel insecticides are subject to the same stringent regulations as other pesticides and their eventual approval must be exercised.It is concluded that although microbial insecticides show considerable potential for the protection of stored products, further research in the scenario is needed.

Now, the molecular tools are available to characterize individual isolates readily and to probe pathogen population structure, and incorporation of such studies into microbial control programmes to facilitate pre- and post-release monitoring that is strongly encouraged. Policies are needed to be amended to permit greater inclusion of integrated pest management tactics in the grain inspection and marketing systems.

REFERENCES

Alaa, A.; Hamdani, A.l.; Merri, M.; Mohamed, K. Effect of bacterial toxin for bacteria, *Bacillus thuringiensis* (Berliner) In some life stages of grain beetle (Kahpra) *Trogoderma granarium* (Everts) (Coleoptera: Dermestidae). *Biochemical and Cellular Archives*. 2018, 18: 2293-2297.

Athanassiou, C.G.; Kavallieratos, N.G.; Evergetis, E.; Katsoula, A.M.; Haroutounian, S.A. Insecticidal efficacy of silica gel with *Juniperus oxycedrus* ssp. *oxycedrus* (Pinales: Cupressaceae) essential oil against *Sitophilus oryzae* (Coleoptera: Curculionidae) and Tribolium confusum (Coleoptera: Tenebrionidae). *J. Econ. Entomol*. 2013, 206: 1902-1910.

Athanassiou, C.G.; Kavallieratos, N.C.; Menti, H.; Karanastasi, E. Mortality of four stored product pests in stored wheat when exposed to doses of three entomopathogenic nematodes. *Journal of Economic Entomology*. 2010, 103: 977-984.

Athanassiou, C.G.; Palyvos, N.E.; Kakoull-Duarte, T. Insecticidal effect of *Steinernema feltiae* (Filipjev) (Nematoda: Steinernematidae) against *Tribolium confusum* du Val (Coleoptera: Tenebrionidae) and *Ephestia kuehniella* (Zeller) (Lepidoptera: Pyralidae) in stored wheat. *Journal of Stored Products Research*. 2008, 44: 52-57.

Beeman, R.W.; Speirs, W.E. Toxicity, persistence and antagonism of avermectin B1 against stored-product insects. In: Proceedings of the Third International Working Conference on Stored Product Entomology, Manhattan, US. 1984, pp. 246-255.

Brooks, W.M. Entomogenous Protozoa. In: *Handbook of Natural Pesticides, Microbial Insecticides, Part A, Entomogenous Protozoa and Fungi*. Ignoffo, C.M. and Mandava, N.B. (eds). US: CRC Press. 1988, vol. 5.

Canhilal, R. The use of entomopathogens in the controlling of insect pests of stored product. *Scientific Papers. Series A. Agronomy*. 2016, 59: 235-240.

Chiang, A.S.; Yen, D.F.; Peng, W.K. Defence reaction of midgut epithelial cells in the rice moth larvae *Corcyra cephalonica* infected with *Bacillus thuringiensis*. *Journal of Invertebrate Pathology*. 1986, 47(3): 333-39.

Cory, J.S.;.Franklin, M.T. Evolution and the microbial control of insects. *Evol. Appl.* 2012, 5(5): 455–469.

Cowan, D.K.; Vail, P.V.; Kok-Yokomi, M.L.; Schreiber, F.C. Formulation of granulosis virus of Plodiainterpunctella (Hubner) (Lepidoptera: Pyralidae): efficacy, persistence and influence on oviposition and larval survival. *Journal of Economic Entomology*.1986, 79: 1085-1090.

Cox, P.D.; Wilking, D.R. The potential use of biological control of pests in stored grain. Research Review 36, Home-Grown Cereals Authority, London. 1996.

Dales, M.J. Controlling Insect Pests of Stored Products using Insect Growth Regulators and Insecticides of Microbial Origin. Natural Resources Institute Bulletin 64. Chatham, UK: Natural Resources Institute. 1994.

Daoust, R.A. Commercialization of bacterial insecticides. In: Proceedings and Abstracts of the Fifth International Colloquium on Invertebrate Pathology and Microbial Control, Adelaide, Australia, August 1990. pp. 7-11.

Falcon, L.A. Innovative approaches to the industrialization of microbial insecticides. In: Pesticides and Alternatives: Innovative Chemical and Biological Approaches to Pest Control. Proceedings of an International Conference, Crete, Greece, 1990; pp. 109-116.

Fayyaz, S.; Javed, S. Laboratory Evaluation of Seven Pakistani Strains of Entomopathogenic Nematodes against a Stored Grain Insect Pest, Pulse beetle *Callosobruchus chinensis* (L.). *Journal of Nematology*. 2009, 41: 255-260.

Fields, P.G. The cold-hardiness of *Cryptolestes ferrugineus* and the use of ice nucleation-active bacteria as a cold synergist. In: Proceedings of the Fifth International Working Conference on Stored Product Protection, Bordeaux, France, September 1990. pp. 1183-1192.

Frydocva, B.; Verner, P.H.; Bartos, J. The comparative effectiveness of two biopreparations based on the entomogenous fungus *Beauveria bassiana*, especially their effects on storage pests. *Agronomicka. A (Rostlinna vyroba)*, 1989 (50): 247-264.

Henry, J.E. (1981). Natural and applied control of insects by Protozoa. *Annual Review of Entomology*, 26: 49-73.

Iatrou, S.A.; Kavallieratos, N.G.; Palyvos, N.E.; Buchelos, C.T.; Tomanović, S. Acaricidal effect of different diatomaceous earth formulations against *Tyrophagus putrescentiae* (Astigmata: Acaridae) on stored wheat. *J. Econ. Entomol.* 2010, 103: 190-196.

Jassim, H.K.; Abdullah, L.M.; Abd-Al-Ahad, I. Determination of the exact concentration of Beauveria bassiana (Vuill.) to control the larvae of the fig moth, *Ephestia cautella* (Walk.) on stored dates in Iraq. *Arab Journal of Plant Protection*. 1988, 6(1): 44-45.

Kavallieratos, N.G.; Athanassiou, C.G.; Aountala, M.M.; Kontodimas, D.C. Evaluation of the entomopathogenic fungi *Beauveria bassiana, Metarhizium anisopliae* and *Isaria fumosorosea* for control of *Sitophilus oryzae*. *J. Food Prot.* 2014, 77: 87-93.

Khan, A.R. and Selman, B.J. On the mortality of *Tribolium castaneum* adults treated sublethally as larvae with pirimiphos-methyl, Nosemawhitei and pirimiphos-methyl + N. whitei doses. *Entomophaga*. 1988, 33(3): 377-380.

Kirschbaum, J.B. Potential implications of genetic engineering and other biotechnologies to insect control. *Annual Review of Entomology*. 1985, 30: 51 -70.

Kumari, S.M.; Neelgund, Y.F. Preliminary infectivity tests using six bacterial formulations against the red flour beetle, *Tribolium castaneum*. *Journal of Invertebrate Pathology*. 1985, 46(2): 198-199.

Lasota, J.A.; Dybas, R.A. Avermectins, a novel class of compounds: implications for use in arthropod pest control. *Annual Review of Entomology*. 1991, 36: 91-117.

Malaikozhundan B.Vinodhini J. Biological control of the Pulse beetle, *Callosobruchus maculatus* in stored grains using the entomopathogenic bacteria, *Bacillus thuringiensis. Microbial Pathogenesis*. 2018, 114: 139-146.

Mbata, G.N., Shapiro-Ilan, D.I. Laboratory evaluation of virulence of heterorhabditid nematodes to *Plodia interpunctella* Hübner (Lepidoptera: Pyralidae). *Environmental Entomology*. 2005, 34: 676-682.

McGaughey, W.H. *Bacillus thuringiensis*: A critical review. In: Proceedings of the Fourth International Working Conference on Stored Product Pests, Tel-Aviv, Israel, September 1987. 1987; pp. 14-23.

Purrini, K.; Keil, H. *Ascogregarina bostrichidorumn* sp. (Lecudinidae, Eugregarinida), a new gregarine parasitizing the larger grain borer, *Prostephanus truncatus* Horn (1878) (Bostrichidae, Coleoptera). *Archiv fur Protistenkunde*, 1989, 137: 165-171.

Ramazan, C. The Use of entomopathogens in the controlling of insect pests of stored product. *Scientific Papers. Series A. Agronomy*. 2016, LIX: 235-240.

Ramos-Rodriguez, O.; Campbell, J.F.; Ramaswamy, S.B. Pathogenicity of three species of entomopathogenic nematodes to some major stored-product insect pests. *Journal of Stored Products Research*. 2006; 42: 241-252.

Reinecke, P.; Andersch, W.; Stenzel, K. Bl0 1020, a new microbial insecticide for use in horticultural crops.In Proceedings of Brighton Crop Protection Conference on Pests and Diseases, Brighton, UK, November 1990. pp. 49-54.

Rodrigues, C..Pratissoli, D. Pathogenicity of *Beauveria brongniartii* (Sacc. Petch) and *Metarhizium anisopliae* (Moss Sorok) and its effect on the corn weevil and bean beetle. *Anais da Sociedade Entomologica do Brasil.* 1990, 19(2): 301-306.

Sabbour, M.M.; Abd-El-Aziz, S.E. Efficacy of some bioinsecticides against *Bruchidius incarnatus* (Boh.) (Coleoptera: Bruchidae) infestation during storage. *J. Plant Prot. Res.* 2010; 50: 25-31.

Sabbour, M. Integrated Protection of Stored Products. Efficacy of some microbial control agents and inorganic insecticides against red flour beetle *Tribolium castaneum* and confused flour beetle, *Tribolium confusum* (Coleoptera: Tenebrionidae). *IOBC-WPRS Bulletin.* 2014, 98: 193-201.

Sabbour, M.M.; Abd-El-Aziz, S.E.; Sherif, M.A. Efficacy of three entomopathogenic fungi alone or in combination with diatomaceous earth modifications for the control of three pyralid moths in stored grains. *J. Plant Prot. Res.* 2012; 52: 359-363.

Samples, J.R.; Buettner, H. Corneal ulcer caused by a biological insecticide, *Bacillus thuringiensis*. *American Journal of Ophthalmology.* 1983, 95(2): 258-260.

Sarwar, M. Stored grain and stored product mites from Pakistan and Azad Kashmir. *Pakistan & Gulf Economist.* 2004, XXIII (10): 30-31

Sarwar, M. Development and boosting of integrated insect pests management in stored grains. *Journal of Agriculture and Allied Sciences,* 2013; 2(4): 16-20.

Sarwar, M. Extermination of insect pests (Coleoptera: Bruchidae) and damage of stored pulses by different methods in market. *American Journal of Marketing Research.* 2015a, 1(3): 99-105.

Sarwar, M. Distinguishing and controlling insect pests of stored foods for improving quality and safety. *American Journal of Marketing Research.* 2015b, 1(3): 201-207.

Sarwar, M. Microbial Insecticides - An ecofriendly effective line of attack for insect pests management. *International Journal of Engineering and Advanced Research Technology.* 2015c, 1(2): 4-9.

Sarwar, M. Challenges due to bacterial infections of the honey bees and contributions to manage pest problems. *International Journal of Entomology Research.* 2016a, 1(1): 4-10.

Sarwar, M. Prevalence of multiple viral diseases associated with honey bees colony collapse and control of disorders. *International Journal of Zoology Studies.* 2016b, 1(2): 29-34.

Sarwar, M.A Guide to the honeybee's protozoan nosema disease and treatment recommendations for pest control. *International Journal for Research in Applied Physics.* 2016c, 1(7): 1-11.

Sarwar, M. Fungal diseases of honey bees (Hymenoptera: Apidae) that induce considerable losses to colonies and protocol for treatment. *International Journal of Zoology Studies.* 2016d, 1(1):8-13.

Sarwar, M. Multiple Insect Pests Management Options for Chickpea (*Cicer arietinum* L.) Walp Grain With Reference to Storage. In: *Handbook of Chickpeas: Nutritional Value, Health Benefits and Management,* Albert T. Lund and Noah D. Schultz (eds.). Nova Science Publishers, Inc., New York. 2019; pp. 265-302.

Sarwar, M.; Sidra, S.;Mariya, A.; Romana, I. Different nematodes and plasmodiophorids as vectors of plant viruses. In: *Applied Plant Virology: Advances, Detection, and Antiviral Strategies,* L.P. Awasthi (Ed.). Elsevier Inc., London. 2020a; p. 14.

Schulz, F.A.; Laborius, G.A. Strategy for big-integrated control of Prostephanustruncatus (Horn) (Col., Bostrichidae). In: Proceedings of the Fourth International Working Conference on Stored Product Protection, Tel Aviv, Israel, September 1987. pp. 497-503.

Searle, T.; Doberski, J. An investigation of the entomogenous fungus *Beauveria bassiana* (Bals.)Vuill.as a potential biological control agent for *Oryzaephilus surinamensis. Journal of Stored Products Research.* 1984, 20(1): 17-23.

Siddiqui, Q.H. and M. Sarwar.Pre and post-harvest losses in wheat. *Pakistan and Gulf Economist.* 2002, XXI (6): 30-32.

Subramanyam, B.H.; Cutkomp, L.K. Moth control in stored grain and the role of *Bacillus thuringiensis:* an overview. *Residue Reviews.* 1985, 94: 1-47.

Tradan, S.; Vidric, M.; Valic, N. Activity of four entomopathogenic nematodes against young adult of *Sitophilus granarious* (Coleptera: Curculionidae) and *Oryzophilus surinamensis* (Coleoptera: Silvanidae) under laboratory condition. *Plant Disease and Protection.* 2006, 113: 168-173.

Vail, P.V.; Tebbets, J.S. The granulosis virus of Indian meal moth: recent developments. In: Proceedings of the Fifth International Working Conference on Stored Product Protection, Bordeaux, France, September 1991; pp. 1247-1254.

Wakil, W.; Usman, G.M.; Yasin, M. Naturally occurring entomopathogenic fungi infecting stored grain insect species in Punjab, Pakistan. *Journal of Insect Science,* 2013, 14: 182.

Zamani, Z.; Aminaee, M.M.; Khaniki , G.B. Introduction of *Beauveria bassiana* as a biological control agent for *Tribolium castaneum* in Kerman province. *Archives of Phytopathology and Plant Protection.* 2013, 46: 2235-2243.

Index